NOBLE CANINE

Search for the Edge

Jimmie Moore

Steel Crow Productions
Jacksonville, Florida

The material presented in this memoir was framed by old documents, letters, and recollections. I have attempted to recreate as closely as possible what actually occurred, knowing that exact details of specific events and conversations are impossible to recall. In those instances where my memory has faltered, I have taken a few liberties to fill in the blanks without infringing on the story's truth.

Book Design by Sagaponack Books & Design
Cover Design: Hugh Keiser
Thanks to Mary Ann Lynn for copy editing
and Frances Keiser for her guidance

ISBN 978-1-7326719-0-4 (softcover)
ISBN 978-1-7326719-1-1 (hardcover)
ISBN 978-1-7326719-2-8 (e-book)

Library of Congress Catalog Card Number: 2018912748

Summary: A memoir of a young airman who trained and served as a sentry dog handler in the central highlands of Vietnam during the war.

BIO008000 Biography & Autobiography / Military
PET004020 Pets / Dogs / Training
HIS027070 History / Military / Vietnam War
FAM029000 Family & Relationships / Love & Romance

Steel Crow Productions
Jacksonville, Florida

Printed and bound in the United States of America
First Edition

To the nine handlers, my friends:
Vinny, Joe, Barry, Russ, Thomas, Mullin, Bolden, Rejba, and me
For you, we live in history. Noble Canine.

To Vietnam handlers, what-ever, where-ever, whom-ever.
Let your spirit rest.

To my canine brothers and sisters, I am sorry.
Never Again

The Genuine Way

Recollect that the Almighty, who gave the dog to be companion of our pleasures and our toils, hath invested him with a nature noble and incapable of deceit. He forgets neither friend nor foe—remembers, and with accuracy, both benefit and injury.

—Sir Walter Scott, *The Talisman*

Contents

War Waste 1969

by Lynn Skapyak Glenn (Harlin)

Where are the big brave warriors now?
The so-called air pirates, women maimers,
children killers, destroyers of churches,
schools, hospitals, where is he?

His silver, supersonic soarer, his bomb blowing,
truck finding, sea swooping carrier, where is it?
He, the educated engineer, architect, geologist,
economist turned *bon vivant* aviator, where is he?

Four years ago, he flew fearlessly, proud of his mission,
bragging, boasting of the Air medals, the D.F.C.s.
He drank in Cubi, swaggered at Yokuska,
rested in Honolulu. He was proud.
Mom, apple pie and the red, white
and blue were with him.

Now he is in a cell.
He wears pajamas, sleeps on a mat.
He is hot, cold, sick and hungry.
He is bored.
The SAM in the schoolyard,
shot him down.
They dragged him out of the paddy,
moved him through the streets,
filled his ears with taunts,
banged his body with rocks,
tortured him with mind bending,
flesh racking, tired tormenting
questions. Finished with him.
And now he sits, he waits.
He waits for the red, white and blue.

*"War Waste" first appeared in **Time** magazine on Monday, December 7, 1970.*

There were a hundred ways to die in Vietnam, thousands of ways to return maimed for life, but my friends and I came home alive and in one piece thanks to our Sentry Dogs; this is my story.

Chapter One

We Need An Edge

Lackland Air Force Base, San Antonio, Texas – September 1968

At nineteen, I didn't feel like I had much of a choice. I could wait for the draft notice to come in the mail, in which case I'd go straight from Basic Training to the front lines in Vietnam. Or I could enlist and have some opportunity to choose my role in the war.

I wanted to go to computer school, so I enlisted in the Air Force, having made a high enough grade, and finished Basic Training in early September 1968. We recruits hungered for choice assignments. Instead, a lot of us, the leftovers, were told we had the choice of being a cop or a cook. Feeding and protecting is a motherly trait. Cops and Cooks are counted as what?

Whether I'll be cooking eggs 50 different ways or standing guard duty in a bunker in Vietnam, neither is appealing to me, Airman Jimmie Moore. Washing dishes sounds worse, so I'm choosing Security Police (SP) School.

In five short weeks, I would be an Air Force security policeman and on my way to Vietnam, anyway.

After basic training, I had started Security Police School, but early in the program I had to take a week of leave due to the death of my girlfriend Lillian—she was my fiancé—back home in the small town of Starke, Florida. By the time I got back, I had missed too many classes to catch up, so I started Security Police School over. Monday of the final week in Security Police School, our instructor made an interesting announcement.

"Wednesday afternoon," he said, "some instructors from the dog school will stop by your dayroom to recruit volunteers for sentry dog training. It's open to anyone in your class who may be interested."

I leaned over the table and looked at my two friends, Vinny DeGuilio and Joe Barbarise, sitting to my right.

I met Vinny and Joe the first day of my second go around at Security Police School. They were old friends from Secaucus, New Jersey. I was angry, bitter, and pissed off at the world, no fun to be around, but every chance they got they dragged me to the Airmen's Club or into San Antonio to drink beer and dance. They kept me laughing.

"Having a dog could be a good idea," I said, "at least there'd be someone to talk with on guard duty." Vinny shrugged. Joe headed for the door lighting a cigarette.

During noontime chow, we talked about the K-9 instructor's news.

"I like the idea of having a dog," I said.

"I-I d-don't know, man," Joe said. He looked at Vinny over his big-rimmed glasses. "What'd you think?"

Joe was tall, olive skinned, brown-eyed, with kinky (not really) black hair, and slender as a two-by-four. He smoked too much and chugged Pepsi like it was water. He loved the ladies and was always fun to be around, the life of the party in spite of his stutter.

I made the mistake of laughing at the way his comment came out, and the look he gave me could have killed.

"I never heard a stutterer before, so it was funny to me. Sorry, Joe, it won't happen again," I said.

"We always had dogs growing up," Vinny said shrugging his shoulders.

Vinny had green eyes and light olive skin, he stood out in a crowd. The girls in the Women's Air Force (WAF) liked him. That made it so much easier to meet women since I didn't have to do it alone.

"You ever have a dog, Jimmie?" Joe asked. He pulled a bottle opener out of his pants, opened a Pepsi and downed half of it in one swallow.

"All my life." I sat back in my chair and crossed my arms as if to say I knew a thing or two about dogs.

"Did you have one of them huntin' dogs down south chasin' 'possums?" he asked with a drawl. Joe loved to tease me about being southern.

"What are you getting at?" I asked.

"This is different." he said, "These dogs are trained to attack and kill people. Like us." He took one more swig, finishing his Pepsi. "Shit. This is just so different. I'm trying to get my head around it."

"I'm going to hear what they have to say," I said.

Vinny started tapping the table with two pencils. He and I shared a passion for playing drums.

Vinny and Joe exchanged glances at each other. Vinny sat back in his chair and crossed his long legs. Joe lit a cigarette. It was like a signal that they agreed or disagreed on something.

"We'll go, too," Vinny said.

Security police graduation was scheduled for Friday morning. After that we had AZR, Asian Zone Readiness, training for a week. Then we'd get our assignments and orders to ship out after a 30-day leave. We were eager to hear of any way we could postpone our fate.

At least I'll have a chance to go home and say goodbye to my family and friends. Thinking about shipping out to Vietnam makes me wish I had considered cook school. They need cooks in Vietnam, but maybe I could train in New York or in France at one of those fancy cooking schools in Paris and cook for a General. Dream on, fool.

Staying Alive

Wednesday afternoon, the sentry dog trainers walked into our barracks' dayroom. Joe, Vinny and I were there early and made ourselves comfortable in the easy chairs. It was hot, crowded, and noisy.

"A sentry dog can get your asses home alive," shouted a trainer. He had a commanding presence in the room.

"Not one single air base in Vietnam has been overrun by the enemy. Guess what? Sentry dog teams are the first line of defense. That means you and your sentry dog are the first to know about any enemy action. Every SOB in his bed at that base sleeps soundly because dog teams are on watch. How's that grab you?"

We were riveted.

"Dog teams go where others fear to tread."

I couldn't believe the crap I was hearing, people in this man's Air Force who loved what they did. I looked at Joe and Vinny for a reaction. Joe was leaning forward in his seat sipping a Pepsi through a straw. Vinny stroked an invisible goatee.

Good. They're interested.

"We encourage security police graduates to become sentry dog handlers," another trainer said. "Your job will be to patrol the perimeter of the air base you're assigned to with your dog to keep the enemy away from our aircraft and weapons. You get eight weeks of extensive combat and obedience training with a dog here at Lackland and at nearby Camp Medina."

He paused for a few seconds.

"You only patrol at night. That's when the base is most vulnerable. The dogs are your eyes and ears. Sentry dog handlers are the elite section of the security police force.

I felt like a 500-pound bomb was dropped on me. *Elite? Me? Us? Are you kidding?* I never thought about being elite anything. Even when I made Eagle Scout I never gave it much consideration. And, it sounded like having a dog meant no bunkers. I did not want to get stuck in a bunker. But the best part was there would be more delay before shipping out. I would do anything to keep from shipping out. Maybe the war would be over by then.

"Something else you need to know," he said. "Tax paying citizens of the United States of America pay a lot of money for all this training and expect a lot in return."

"Question?" He invited an Airman in the back of the room who raised his hand to speak.

"Sir, why use dogs? Why not just put more men out on patrol or put up more guard towers?"

"The Vietcong are sneakier than Indians. They'll grease themselves up and slip through the night, the fences, the minefields, and you'll have no way of knowing if they are out there or not. They'll crawl through rows of Constantine wire and never be detected by humans. Because a sentry dog's nose is 10,000 times more sensitive than ours, he can pick up a scent over a mile away. And he'll hear something long before you do. When you're on patrol, you depend on your dog's superior nose and ears for alerts. If you command your sentry dog to attack, he can break a man's arm, or take a large chunk from his throat with powerful jaws in the blink of an eye."

I looked at Joe and Vinny. "We gotta do this!"

The K-9 instructors went on for a while longer sharing their experiences and answering questions. They had all been to Vietnam or somewhere else in Southeast Asia.

"Another thing," said a different trainer, wrapping up.

"If I have to go back to Vietnam, give me a dog. They're the first to know that something's going on. You will not see, smell or hear before they do. That's the edge. Sentry dogs are your edge."

The edge; these dogs are our edge.

The instructors invited us to drop by the kennels over the next couple of days to see the dogs and ask questions. If interested, we could sign up at the kennel office.

After they left, we went back to Joe and Vinny's room to talk it over. Another selling point for me, besides being with a dog and delaying shipping out, was more schooling. Having little combat training in Security Police School and with only limited time on the firing range during basic, none of us was ready for war. We didn't have a clue.

Joe, Vinny and I needed an edge and these dogs would help. With advanced education and training, more young men like us might come home alive from this war.

"You really want to sign up for this?" Joe asked me.

"Let's at least check it out," I said.

I was trying to be calm and cool about it and not let them see how excited I was. Every day since starting Security Police School the barking from the kennels was always present. The constant barking never bothered me and I didn't give it much thought. I even took a walk over to the kennels on a lunch break one day just to see these dogs up close.

Walking among the dogs was the strangest feeling, like they were wolves wanting to attack and tear me limb from limb. All these dogs wanted was for a chain to break and I would be running for my life. The coolest thing to happen was when someone shouted, "Shut up!" and they did shut up, not for very long, but the seconds of total silence just blew me away.

I'm hooked.

The kennel area reminded me of our county fairs with the 4-H and all the livestock. The smell of fresh hay and manure filled the air. I watched some handlers from a distance working with their dogs. The way they groomed them reminded me of 4-H-ers after showing their livestock at the fair back home.

I remember watching a young girl from 4-H crying as she groomed her first-place Angus steer. She knew he was headed for auction, which meant the butcher block, but it made no difference. For years, she'd cared for, raised this calf and loved him, but she would let go…had to. That's what farmers do.

How can I convince Vinny and Joe to come with me?

Back in class that afternoon I couldn't contain myself as I blurted out to my friends. "Everyone will think we're bad asses 'cause we're elite!"

So much for staying calm and cool.

Thursday afternoon, the three of us went into the kennel area together. I had observed the kennels the evening before, but now it was up close and personal with my two friends. There were maybe a few hundred German Shepherds of every shape, size and color, and all of them were running around their birdhouses barking, growling and snapping like rabid hounds from hell.

A birdhouse was a white doghouse that sat on a ten-inch wide pole about two feet off the ground in a ten-foot square covered area. It was funny seeing dogs on top of them. How they got on top was a mystery. Dogs were secured to the pole with a six-foot long solid steel chain, giving them 360 degrees to run, but no fence to keep them in if they broke loose. There was also an enclosed building nearby full of regular kennels.

It was easy to daydream about where all the dogs came from and wonder about the families who had given up their pets.

"Hey Vinny, why would a family want to let go of their pets knowing their dogs will be used as military working dogs? Don't they realize that their dogs will be going to Vietnam? That they'll be working perimeter duty as the first line of defense at an Air Force base where they could be hurt or killed?" I asked.

"Maybe the recruiter lied to them, just like he did me," Vinny complained. "Why, your dog will be stationed in the South of France or in Germany, depending on the needs of the Air Force, you can count on it," he'd said.

"Rat crap" I shot back.

What about that young 4-H girl and her bull? It's what farmers do, and it's what a patriot does also.

The first dog we saw was Bart, a huge, shaggy, black beast of a German Shepard trying to break free with all his 90-pound might. We were told no handler had been able to control or work this dog. The devil's own dog couldn't be meaner. If that dog broke loose, we knew we were lunchmeat. All the dogs were nuts, running around their doghouses barking, salivating and gunning for us. It was so overwhelming. And it felt so right.

"Are you shittin' me?" Joe whispered.

"Isn't it great?" I said.

"It's awesome." Joe grinned back at me.

The dogs were getting worked up. They were straining hard against their chains. All I saw was teeth, lots of big teeth.

It was easy to tell when a dog liked someone and when he didn't. Most of the time, a dog would just sit and look at a handler, maybe scratch something. Some sentry dogs liked an Airman right off, which was made clear by their tails moving at lightning speed or they had that doggie 'grin' on their faces. It was a K-9's way of saying, "I'm OK now. You can take me out for a walk. Promise not to bite you, you can trust me, really."

"Would you look at all these dogs?" Vinny asked. "Every damn one of them wants to tear our heads off."

I laughed and said again, "Isn't it great?" I couldn't help myself. It was like poking them in the ribs. *Come on guys, let's do this.*

They were frightful puppies, but I knew we could take advantage of all that fierceness. We crept through the birdhouses. The barking grew more intense. I guess at least several hundred dogs never took their eyes off us. It kind of grew on me but it was creepy. We made our way to a smoking area to get away from the noise, to a spot where we could talk.

Joe was first to speak as he let a long pull of his cigarette escape and mix with the kennel smells that filled the air. "More and more I can see the edge in these beasts. I can see 'em, hear 'em, and damn sure can smell 'em."

We laughed as we took in the kennel area where, to our surprise, most of the dogs had stopped barking.

"We know we're going to Vietnam and I don't know about you all, but I need to know my options. Guys, dog school is an option, is it not?" I said. I was getting anxious.

I started pacing, with arms and hands flying through the air. My Italian friends' gesturing had worn off on me.

"These K-9 instructors are the only ones suggesting any alternatives against some pretty crazy odds," Vinny said. Now he was up and pacing, too.

"At least it's safer out front where we can move around and have a trained attack dog at our side. I get it, I get it." Vinny was on a roll. "And it's the edge the Sergeant was talking about."

Vinny was as white as a ghost as he looked around and took it all in. He shrugged and looked at Joe, eyebrows raised. Joe was laughing at the both of us. We looked like two strutting roosters talking like dueling preachers on a Sunday morning.

"OK, I'm in," he said. "I think this is our edge, these four-legged maniacs."

"You think we've got a choice?" I said. "Look guys, we're one-stripers. We're nobodies. We have a choice of a bunker or maybe a tower, if we're lucky. And ten bucks says you gotta do some serious ass-kissing to get in a tower."

"Jimmie, you're talking out your ass, man, not one of us has a clue what is going on over there," Vinny said.

"Those cops getting killed in that bunker makes me want to be in a tower," Joe said, "I'll kiss whoever's ass."

"No, you won't. I know you better than that and you're no ass kisser," Vinny said. "What bunker you talking about?"

"That big air base in the capital city, I don't remember what the Sergeant called it," I said. "Nine security policemen died during Tet last year, the Vietnamese New Year. Sappers strapped down with explosives, broke through the base perimeter at Saigon's Tan Son Nhut Air Base and blew up their bunker."

Joe interrupted. "What's a sapper?" he stuttered.

"Come on man, it's a Vietcong commando," Vinny said. "Remember, Sergeant told us in one of our classes that the Air Force never repaired the bunker, just left it as a reminder."

"I'll bet those security policemen didn't let their guard down at all, the Vietcong got lucky or maybe they are just damn good at what they do," I said.

Joe and I stood in stunned silence as Vinny took a deep breath and leaned against a picnic table.

"I'm scared of heights," I said. "This is easy for me and don't change the subject. The reason they got their asses blown away was simple, no K-9 out front. I'm going to dog school."

"Not one of them dogs is happy to see us," Joe said, laughing. He drew on the cigarette he had just lit, inhaled and exhaled, blowing smoke through his mouth and his nostrils.

I was frustrated with these two. I stopped pacing and looked both of them square in the eyes. "A lot of American boys just like us are dying over there, and in fact, most of the ones dying are exactly like us," I said.

I pounded on my chest. "I don't want to see the inside of a body bag." That outburst got the dogs going again.

"SHUT UP!" I yelled at the top of my lungs. But the dogs got louder and louder. We could barely hear ourselves think much less hear each other over all the noise. We stood there and took in the scene.

Vinny, Mr. Quiet and Thoughtful, was shifting from one foot to the other. Yelling over the noise, he argued how the dog teams were the first line of defense and the first to be introduced to Mr. Charlie.

"Well, no shit, Sherlock," Joe said.

"We'd be outside the fences, in front of trigger happy guards in bunkers or towers, doesn't matter which," Vinny said. "Those bunker guys get scared, they might blow away both dog and handler. Killed in action by your own people, that's some scary shit."

Joe threw up his hands. "But has it really happened, bunker guys shooting K-9 guys?"

"It's simple," I said. "You can sit in a bunker with the rats or get out, move around, hide and protect yourselves better with a dog. We'd be the hunters instead of the hunted."

"What rats?" said Joe.

"Didn't the Sergeant say the bunkers had rats in them?" I said. "I'll take a wet nose and pointy ears any day or night over being hunkered down in some bunker with no place to go."

I threw up my hands mocking them both. We acted as though we were trying to convince each other, but down deep it was a no brainer. We all knew it. We didn't have any options except go to dog school.

"Look," I said, "you got this dog that can smell 10,000 times better than you can and that's what we want, right? The dogs alert on the enemy first. We need that edge in this man's screw up war."

Joe and Vinny looked at each other. Joe took off his glasses and wiped the sweat off his face. He sighed and lit another cigarette.

"I'm in," Vinny said. "Let's go eat, all this talking and thinking has made me hungry."

"And Vinny, you get eight more weeks to get close to that gorgeous WAF you've been chasing," I said. He shrugged his shoulders, smiled, and nodded his head in agreement.

Vinny summed it up. "I know one thing, we are all coming home from this war with all our limbs and our minds. No one is going to die and if having a dog gives us an edge, damn right I'm going to dog school."

We signed up at the kennel office. Thirty were allowed in the class and we filled it. Our orders should be changed and we'd move to the K-9 barracks right after graduation from Security Police School. We were going to dog school. Hot damn!

"Let's eat," said Joe.

The Single Life

"You know, guys, I think I've figured out something," said Vinny.

We were hanging around in the barracks dayroom after dinner that night, shooting the breeze.

"Oh, good grief, what have you figured out now Vinny?" said Barry Teagan. Barry was the most likable of all in our little group. He was a good humored, big ole teddy bear from Michigan who loved to eat, so he was a little overweight but carried himself easy enough.

"Look, I'm serious as a heart attack," Vinny said, "I've really come across valuable information that I could easily sell."

"Here at Lackland Air Force Base all valuable information belongs to the Air Force," I said, "but please, pray tell, what is this information that everyone will pay to have?"

"WAF do not like short hair on men, even though both wear the uniform," said Vinny.

The ladies of the WAF understood we were going off to war and may never get the chance to kiss anyone again. Some of these women would stay here at Lackland and that's where Vinny, Joe, me, and lots of other guys stepped up to help make these women as comfortable as possible.

"Brilliant, Einstein," said Joe.

"Really, Vinny, that's it? Long hair is the key to winning the hearts of Air Force women?" said Barry. "Seriously?"

"He's right," I said.

"Here we go," interrupted Joe.

I shot back. "No, listen brainsless, not a one of us has seen the inside of a barber shop since basic training, and I'll bet money on that."

"It's a fact. If you want to date a pretty lady here, first, stop getting your hair cut and get some clothes," said Vinny. "We are not Marines. Second, buy nice clothes, a pair of shoes, as much as you can afford."

First thing Vinny did was find an Italian shoe store. He tried to get me to buy a pair. "Look," he said, "These shoes are so light you can put them in your back pocket."

And he did, but, just the same, I wore the nice ones the Air Force issued to all of us. Shoes are shoes to me, not so Vinny, the one who had the class in the group. Joe was the same, they both bought nice black slacks but Joe passed on the shoes. We couldn't buy much on $200 a month.

I stuck with Levi's and a nice polo shirt. I invested in some tennis shoes after slipping on the sidewalk in the rain because of the slick leather soles on my uniform dress shoes.

Vinny and I had new clothes, shoes, and hair. We were ready.

"Joe, like, where do you go for the weekends?" Vinny asked.

"Yeah! Why do you not hang out with us?" I wanted to know what he had going on, too. He wasn't talking and we didn't push it.

Dating was different now, we were men with grown men's desires, but I didn't have a clue about the game or the chase. That's why hanging out with Vinny made being around women easier. Did I mention that Vinny and girls were like bears and honey?

Graduating from Security Police School was a dull non-event. There were thirty in our class which brought about twenty-five or so family members, wives and girlfriends.

Our uninspiring Squadron Commander gave a speech about "how important your roles are in protecting life and property." Wherever our next assignments would be, anywhere in the world, our "value as protectors would not go unchallenged, but Security Policemen endeavor to persevere," said our no-name Commander.

We were glad to be finished with Security Police School even though there was no one waiting around to see us off. There were some parents in attendance for our graduation along with a few girlfriends or wives, but as for Vinny, Joe, and myself no one arrived to congratulate us.

We finished packing and made the short trip of about 100 yards to our new barracks, got checked in, unpacked, and settled. In less than an hour, we went from one school to another.

On Friday, Vinny, me and our WAF friends were going on a date, our first date since arriving at Lackland AFB. We were excited being able to get off the base and going to downtown San Antonio with some ladies hanging on our arms.

My date Mary was a petite, strawberry blonde WAF with green eyes, and she looked fantastic wearing grey slacks and a form fitting white sweater. Rebecca, Vinny's date, was a knockout brunette in a pink mini-dress. She came from New Orleans.

"Look," said Vinny. "I found an excellent restaurant on the canal with prices we can afford. The food is supposed to be great."

The four of us boarded the bus for San Antonio. Public transportation played a key role for budding romances in the Air Force.

On the way, Rebecca said, "This will be a Dutch treat date, guys?"

I had money and so did Vinny.

"Sure," said Vinny. "Going Dutch is just fine with us. Maybe that way we can afford a beer at the enlisted club next weekend."

After dinner, we went to the movie theater. "Hey, let's get a big tub of popcorn to share," I said. The girls agreed, though we barely touched it except for a couple of dunks into our Coca-Colas. Great stuff.

I can't remember the name of the movie or who the actors were and we didn't care. On our return trip, the bus driver was kind enough to dim the interior lights and we all took advantage of the semi-darkness. Even though touching another girl scared me at first, Mary had me busy making out in the back of the bus and I loved it.

I think I'm getting over my lameness, but why do I keep thinking of Lillian? Can Mary tell how much I miss her? Be still.

On my return to the barracks, I found an already sleeping roommate.

I slept hard and had good dreams of Mary. Red flashing lights and a broken car replaced the good with the ugly nightmare about my fiancé, Lillian, who had died as a result of a train collision.

On that night, three young women, one of them Lillian, had gone to a movie and later some supper. While taking Janet home they came to an old railroad crossing with lights flashing and a train flying by. They stopped, not thinking anything about it except that's what people do when a train is coming through.

Once the train passed, they crossed the tracks and dropped Janet off at her home about fifty yards beyond, then returned to the same railroad crossing where the same lights were still flashing, but no crossing bar stopped them from driving onto the tracks to collide with another freight train that was travelling at fifty miles per hour. Gloria, the driver, died instantly.

Within hours of the accident the Red Cross went into action, and I was excused from my security police training to fly home to be with my fiancé. My Dad had no new information to tell me as we made the long journey to Gainesville and Shand's Teaching Hospital. The nurse at the front desk informed me that Lillian had died from injuries to her head just after four a.m. I had been in the air when she died.

When I awoke, my roommate had already split. I arose, cleaned up and walked outside into a picture postcard of a cold Texas day. This Saturday morning was the kind of fall day when we still thought about the beach and lying in the sun. It didn't take long for the shirts to come off, the music to get loud, and cold beer improved the situation. It was going to be a beautiful day.

I found a spot to lie down and made myself at home. It didn't take long for the effects of the warming sun to make my eye lids heavy.

A Marine Has Landed

"Hey! Lookie here, we got ourselves a Marine," said one of the sun worshipers.

As soon as he stepped off the bus, I knew I was looking at a Marine. He could have easily been the one pictured on the billboard outside the recruiting station. As he walked up the slight hill to the K-9 barracks, it was like he was on parade. His stride was almost a march. He knew he had an audience, and why wouldn't he? This guy was a picture of stout manhood and everyone just knew they would be safe in his foxhole.

Foxhole? Please Lord, forbid it! I joined the Air Force to avoid foxholes.

He was at least six feet, my height, and looked to be a lean mean fightin' machine. And he was handsome in a pretty boy way, with blue on blue eyes, blonde hair and a strong chin.

I still had some baby fat, but was in good shape after six weeks of basic training. I also knew that running a race against him would be a waste of time.

He looked sharp as a tack and wore the uniform well. In comparison, I found our Air Force blues lacking. We looked like elevator operators you'd see in old movies before sound was introduced. The Marines knew how to dress up their men. This man wasn't in his dress uniform but he still looked great. The Air Force could learn something about the one-size-fits-all uniform we wore.

When he got to the entrance, he stopped and stared at us.

"Which way to get checked in?" He sneered.

Great, we have ourselves a Clint Eastwood.

"First floor dayroom," someone said. "Hey! Are you really a Marine?"

He paid no attention and disappeared inside.

It was a fine day to be alive and the rays of the sun were like a narcotic. I drifted between melancholy dreams of my hometown girl and sweet memories of my date from last night.

Lillian, oh Lillian, what do I say? How do I tell the world of our love and how special a young woman you were?

Something was biting my foot, bringing me out of my dreams. Joe and Vinny were standing over me.

"Wake your ass up. We're headed for some chow," Joe said.

"I'm hungry, let's do it." I said.

As we grabbed our trays and headed for a table, I noticed the Marine sitting alone, so I invited us to join him.

"Are you the Marine that came in this morning?" I asked.

"Yes," he said. He jammed a large meatball in his mouth.

"Well, welcome to the Air Force. Where are you from?" I asked.

"New Mexico, Santa Fe," he said. He never took his eyes off his food or slowed down to talk. We just looked at each other.

"How's Air Force food?" Vinny asked.

"Damn good," he answered. He jumped up and headed to the chow line for seconds.

"A man of few words," said Joe. We dug into our chow.

All four of us ate in silence, enjoying the food.

As the Marine finished up and sat back in his chair to relax, I took the opportunity to ask another question. "Where do you suppose you'll be stationed as a sentry dog handler?"

"Have no idea, but I hope to go to Vietnam," he said.

"So, Vietnam is something you want to do?" Vinny asked.

"Well, hell yeah! Don't you?" he said, "What else does a Marine live for but to fight?"

We were in the presence of a gung-ho Marine.

"Hell no, I don't," I said. The death of my fiancé had given me an attitude that hung on me like a wet bathing suit full of sand. "That's why I joined. Now, the Air Force is sending me anyway. At least I had the good sense to volunteer for dog school."

"Chicken shit Air Force," he said. He got up and left.

Joe swore under his breath. "Well, screw you very much."

Later that evening, I had my second date in as many days with Mary, this time just the two of us. Holding her in my arms as we slow-danced was heaven.

When it was time for a break, I ordered two beers and we sat down at a table to chat. "I love beer. Is that trashy?" she said. "I can't help it. I grew up sipping home brew that my older brothers made. My mother would have killed me, and my father would have done it all over again had they caught on."

Since we were revealing our secrets, I took a risk and brought up Lillian.

Mary wasn't put off at all. "How did you meet?" she said.

"Lillian and I discovered each other walking across the high school bus parking lot," I said. "Our eyes met, and in a flash my heart was hers."

"How romantic!" Mary's green eyes crinkled when she smiled.

"By the weekend we had our first date, a double date with a close friend and neighbor. A movie and a late-night hamburger made Lillian late, but a phone call to Mom bought us some time for a first kiss."

I squirmed, thinking it might have been a bad idea to bring up kissing another woman. Mary read my mind. "It's okay, go on."

"From then on," I said, "I saw her every day at school and her parents allowed one night a week for dating."

I was much encouraged and opened up enough to tell the rest of the story—of our three-year courtship, our engagement, how I had lost Lillian to a train, how she haunted my dreams.

By the time I finished, tears were streaming down her cheeks. "Mary, please don't cry," I said. "You're shedding enough tears to make anyone think you knew Lillian."

She sobbed and brushed tears from her face into her soft reddish hair, and I held her, but I couldn't cry. Her kisses that night filled a hole in my heart and drained me of my grief.

Welcome to the Edge

The door opened. The room got quiet as we found our seats and the instructors entered. Four men stood before us looking sharp in their starched green fatigues, wearing their Air Force blue nametags with white lettering standing out on both chest pockets. They were as clean cut as any American Airman could look. Seemed everybody at Lackland AFB dressed in starch.

Of the four instructors' uniforms, Sergeant Vernon Anderson's was the stiffest. He was of average height and build, with brown eyes, dark hair, and reddish-toned skin, and he had a commanding and stern presence that marked him as a leader.

Anderson snapped forward two steps, his uniform scritch-scratching, and began to speak. "Welcome to Sentry Dog Training School. I am Sergeant Vernon Anderson, your lead instructor. These men behind me are your instructor team: Staff Sergeant Adamson, Technical Sergeant Lotes, and Sergeant Gildart."

Sergeant Anderson began morning instruction with K-9 warfare history. Egyptians, Persians, Greeks, Romans, Chinese and Japanese, all had used war dogs. Our morning instructor showed us slides of war dogs in armor, carrying weapons and ammo, as messengers, or as couriers of medical supplies and even food. I learned that Rin Tin

Tin was a real World War II war dog and hero before he went on to make it big in Hollywood. It was Napoleon in Egypt who suggested chaining dogs to the outer wall as a first line of defense warning system. During World War I, the French and German Armies used an estimated 50,000 dogs. The American forces had no dogs but borrowed from the British, French and Belgians.

"Joe, I never heard any of this stuff about dogs before," I said.

"News to me," said Joe.

Dogs were also used in local police actions, which were detailed in the lecture. We already knew about this since we watched the news broadcasts of German Shepherds being used against civil rights protesters in Birmingham, Selma, and other places in the south. The sight of frightened people knocked down by water cannons and bitten by police dogs was not my cup of tea. But, I was a cop.

The Vietnamese are 'gooks', enemies to be hated and killed. That's easy, but our own citizens. That's not right.

"Joe, Vinny, I'll tell you right now, I've made a promise to myself to never turn my dog loose on anybody but law breakers. People fighting for the right to use a damn bathroom are not criminals."

"You wear the badge, you walk the thin blue line," Vinny said, "or something like that. Goes both ways, but being a police officer is all about rules. I'm still a student."

Soon enough I'll be a cop and a dog handler.

The afternoon classroom instructor was the Kennel Master, Technical Sergeant Lotes.

Lotes was a studious-looking beanpole of a man, about six foot one. "I am Technical Sergeant Lotes."

Lotes detailed how records were kept on everything from what the dogs ate and drank to shot records and regular trips to the Vet. There was a resident Veterinarian located at the kennels. With all the health issues a dog could have, it was necessary to have a Vet on base.

When Lotes finished the health segment of his talk, Sergeant Anderson explained that once we got to know our dogs, we would learn to recognize their alerts.

"Most dogs look in the direction of a suspect," he said. "The ears go up, sometimes the dog freezes and stares in the direction of the alert, sometimes the sentry dog starts to move in the direction of the alert. Each dog is different but the results are what counts."

Instructor Adamson was up next. He stood a few inches shorter than Anderson, but his thin, wiry frame was energetic and his movements were precise. His piercing blue eyes made a striking contrast with his pale skin and reddish hair.

"I am Staff Sergeant Tony Adamson. This is a dangerous job, and my high standards for your performance are to keep you safe."

Adamson scanned the room to make sure he had our attention. "First rule: Do not come unprepared for a test. You could die if you are unprepared on guard duty, and you will wish you were dead if you're unprepared here."

A few of us sat up a little straighter in our chairs to show how attentive we were being.

"Second rule," said Anderson. "Never, ever, turn your back on your dog. Now, let's talk about the suit. All of you will be required to put on a padded suit and pretend to be bad guys."

"Better not fall," said Sergeant Lotes. There was a grin on his face. "They love going for the balls or face!"

A few guys thought this was funny, but no one was laughing when Anderson cut us off.

"We are not kidding you," he said. "These dogs know flesh is better than a rag. Any flesh that is not covered by the suit is fair game for one of these passionate dogs to go after."

Sergeant Anderson paused for a moment as if running tapes of old training cycles in his mind's eye, then resumed his talk. "The choicest spot is head level to a dog, your balls. If you do fall, cover your face, close your legs, and hope the dog does not go for your feet."

At the word "balls," I winced. And, so did Barry and a couple of my other classmates I could see from my spot.

"Listen up, men," Adamson said. "I've seen enough blood from being in these padded suits to tell you these dogs like the taste of your flesh and blood and it changes them for the better, I think. It makes them meaner and more aggressive."

Airman Russell Puls gulped. He was a pimply-faced surfer from California, fearless in the ocean, but he was terrified of being bitten by one of these beasts.

Adamson stood. "On Friday, when we issue your dogs, some of you will forget the warning not to turn your back on your dog. We'll see, it happens in every class."

Imagine us writing letters home about being attacked almost daily by nasty-mean, unkind, despicable German Shepherds. "Oh, hello Mom, had a wonderful day, you should have been here. I got attacked several times by a dog weighing a hundred pounds that wanted to eat my balls. Everything turned out OK. I even got to wear a suit. Life is good, Mom. Much Love."

Staff Sergeant Adamson continued. "This may be dangerous training, but for the most part the new handlers love getting into the suit. This is fun, don't make work out of it."

It was unsettling to watch the instructors laugh and wink at each other.

"Listen up, people," Sergeant Anderson said. "Most of your dogs have cycled through two or three times of the eight weeks of training. You've got it easy in that the dog will be training you, but obedience training has to be repetition on top of repetition, and it will be done every day until you leave here. You get lazy and the dogs get lazy. When you get to your first assignment, continuing to do obedience training is a must. Your lives will depend on it."

Anderson made a steeple with his hands. "Right along with more obedience training, your instructors will start the process of adding know how, repetition, patience, praise, correction, drill and more drill along with familiarization, formations, commands, commands from a distance, confidence course, and then off-leash obedience training."

Adamson motioned that he had something to add. "Not all dogs want to be war dogs," he said. "Some dogs are not willing to sit, stay and come. Some dogs are very sensitive to gunfire. That could be a problem in Vietnam so we need to weed them out."

"Sergeant, is there really such a thing as a gun-shy dog in this man's Air Force?" said one of the Army men, "And what happens to the dog if it can't be corrected?"

That got a few laughs from around the room.

"With special training you may be able to save the dog and turn him into a good working dog. It's done all the time. If we are unsuccessful in training this behavior out of the dog, then the dog is removed from the training."

"Sergeant, what happens to the dog if you can't?" I asked.

"Airmen, any dog that cannot be trained is drummed out of the Air Force and sent to the Army or the Marines."

That broke the men up with much laughter, until Adamson barked an instruction.

"Read your manuals, every page, by Thursday," he said. "There will be a short test Friday morning. And be prepared."

"Boo!" Vinny and Joe led the chorus.

"Ok, last words of the day," said Sergeant Anderson. "Read the entire manual by Thursday and that's an order. You're dismissed."

Our Marine friend wasn't pleased with the joke at the expense of the Corps so he got up and left in a huff. The first day of school was over.

A Taste of Hatred

As the three of us started for the chow hall, I told them of an experience I had with two elderly black couples. We were talking about the news broadcast last night with good ole Walter showing us how police dogs were being used against people somewhere, didn't catch where but the deep south for sure, and I began to tell a story that happened to me back in my hometown. I told Vinny and Joe about the incident, a racial incident, a hate crime.

When I was sixteen, I got a summer job working at a car dealership in my hometown. Pumping gas and prepping new cars for new car buyers was my routine. Things slowed down on the weekends and I mostly pumped gas and took care of the driving public. One afternoon a nice- looking Cadillac pulled in for gas. I got the gas started and was checking under the hood as two elderly couples got out and stretched. The ladies approach me for the location of the bathroom.

"Oh shit, I can see it now." Vinny said.

"Were they black?" Joe asked.

"Yes, they were and believe it or not I didn't think a thing about it as I pointed towards the inside of the building. What happened next just blew me away."

"Now I bet the shit hit the fan." Joe said.

"Big time, Joe."

"My boss comes running over and tells the ladies that the bathrooms are broken and closed for repairs."

"Both bathrooms were closed?" Vinny asked. "What happened?"

"They were standing about thirty feet from me and I wasn't quite sure where the conversation was headed, so like a dummy I hollered out that the bathrooms are ok."

"Well, what's wrong with that? Were they working or not?" Joe asked.

"They were working just fine but the problem was my boss had told me not to let any blacks use the restrooms. I really had forgotten but it was so bizarre denying anyone access to a bathroom. It saddened me."

"Excuse the pun, yeah right." Vinny laugh but added, "I can't imagine denying my grandmother or father who are about to bust a gut to relieve themselves."

"The guy who I worked for was recently moved to Starke to take over the dealership. He brought the whole family with him, too. They were from Mississippi and did not like Negroes."

"So, did they get to use the latrine or not?" Vinny asked.

"Even as other customers were pulling up for service and the two couples started causing a scene, the man would not let them use the bathroom."

"Well come on, how did it end?" Joe asks.

"The one black man paid the bill and they all got in their car and left without ever using the bathroom."

"You're kidding?" Vinny said.

"No way man, no way, you are telling me they just up and left, without relieving themselves?"

"That is actually what I am saying."

"As they pulled out I had that same thought, they never relieved themselves. I felt terrible about it and vowed never to lower myself by refusing anybody the right to relieve themselves. I thought about my grandmother being refused access to a bathroom."

"Man, I can't believe this happened," Vince said noticeably upset. "Man, when you got to go, you got to go, period. Don't hand me this crapola about the bathroom being out of service."

"Those two black couples should have dropped their drawers and taken a dump right there in the lobby. That would have been a mess especially with other customers coming on the scene."

"I hear you, Joe, but the story gets better." I was excited now.

"I bet they fired your ass." Vinny volunteered.

"Yep, you be right about that but not on the spot but a few days later. But after the black couples left, my boss tells me that 'the fucking toilets were broken to all blacks.' I asked him about the need to really go, like I really have to shit, like right now. Boss man didn't care, no exception."

"Your boss is a real hard ass when it comes to elderly people." Vinny added.

"Yeah, you mean elderly black people."

"The next week on about Wednesday I was called into the office and told by the owner of the business that I was being let go because a relative was moving to Starke and needed the job. So ended my relationship with the dealership and I forgot all about it until school started."

"What are you saying?" Joe wanted to know. "Are the chickens coming home to roost?"

"You bet. I was pulled out of class one day and told to report to the principal's office. Oh boy, what have I done? I was taken into an office and introduced to two FBI agents."

"Oh my gosh, you're kidding right?"

"No, I'm not. Guess what they wanted to talk about."

"Old black people who needed to facilitate in Starke and the toilets were broken."

"That's right Vinny, you were listening."

"The FBI agents were asking questions about the incident between the black couples and the dealership. They didn't say anything when I told them I had lost my job. Wasn't a big deal and that was the last I heard of it."

"Boy, Jim, you really screwed up," Joe said with a stupid grin on his face.

"Yes, Joe, I know what you are going to say. I needed a good lawyer."

"Damn right, you should have sued the bastards."

"Joe, I didn't think about it and no one around me was thinking about it or even knew about it. In fact, the FBI agents told me their investigation would continue and I was not at liberty to discuss the case with anyone. Not even my parents. Seems these two black couples filed civil rights suits against the owner of the dealership and I ended up being the only witness to verify their side of the story. I walked out of that interview feeling ten feet high and I enjoyed every minute."

Talking as we ate, we decided to form a study group that would meet in our room that night and see if there was some way to get through those hundred plus pages of the manual.

"Hey, should we ask the Marine to join us?" Barry said.

Vinny smirked. "Why not? Surely it can't hurt to extend a helpful hand to a brother handler."

"Sure, we might find out he is human after all," said Barry. We all had a belly laugh.

But the Marine was having nothing to do with any of us or our study group, which was OK by us. We crossed him off our happy list and forgot about him.

The first meeting of our study group was made up of Joe, Vinny, Barry, me, some new guys we just met, and Matthew or Matt. Matt was my new roommate, a short, muscular, energetic guy with dark hair and eyes and a deep tan. Of our group, he was the only one who had a car.

We had just four evenings to get through the hundred and twenty pages, so we decided to break down the manual into parts. On each of the four nights, we would go over two sections. We made good use of yellow marker pens and managed to get nearly halfway through the manual the first night. We would be ahead of our classmates tomorrow.

"Let's go somewhere. Who's up for a ride?" asked Matthew.

It was a tight squeeze but we still managed to get seven of us in Matthew's car. Vinny was in hog heaven because, as it turned out, Matt's car was the same year and make as Vinny's car.

This Chevy's exterior was white with a lot of chrome. The interior was red leather and more chrome. Back in Starke at the dealership, I had seen a similar model as a convertible. It was the coolest car I ever rode in, and to be able to ride around Lackland in one, we felt like kings. It was a pure joy to ride in an automobile again. Being his roommate, I got to ride shotgun most of the time.

Poor Vinny never got behind the wheel, for old Matt had a "no driver but me" policy and he would not budge. Could you blame him? The car was too classy to let just anybody behind the wheel.

On Tuesday, the second day of class, our instructor said, "Look around at your classmates. All of the service branches are represented."

Lackland Air Force Base was where all sentry dogs and their handlers were trained, regardless of military branch. As I looked around, I saw that the Army, Navy and Marines were represented. Well, one Marine, anyway.

"Sergeant, I've a question for you. Where do all these dogs come from? I have never seen so many dogs in one place."

The question came from one of the Army guys and it was a good question, I had thought about it myself.

Technical Sergeant Gildart, the Kennel Master, took the question. He was as big as the Marine, and better looking, the kind of guy the girls in Starke would have swooned over.

"These dogs come from all over the country and most of the owners donate them out of patriotic duty. We buy some from kennels, tried several breeders from Germany, too, but the price tag gets expensive. The Air Force isn't big on breeding their own dogs yet, sort of in the beginning stages, experimental to see if it is cost effective."

Sergeant Anderson paced the floor in his stiff uniform. "Teams travel to all parts of the country and screen each dog," he said. "Half of the dogs will fail this first screening that includes both medical and doggie basic obedience. About twenty percent are rejected due to physical problems, such as hip dysplasia and heartworm disease. Another twenty percent are rejected for temperamental causes, like being gun-shy or under aggressive, even for being too aggressive."

"Question, Airman?"

"Yes, sir," said Russell. "What happens to the rejected dogs?"

"Most of the time the dog would be returned to their families at considerable cost and logistical concern to the Air Force. The sick ones are euthanized. At any given time, we have over 1,600 sentry dogs serving just in Vietnam. Keep in mind that almost half of the animals wash out, so we are always looking for a few good German Shepherds."

I'm looking forward to Friday when we will be assigned our dogs. Some of the guys seem nervous about being bitten. I don't want to think about it.

On Wednesday in class, Sergeant Gildart said, "You're not a dog handler until you are bitten."

That was some real macho crap, I sat thinking.

That night our study group met again. We had grown by one student, Craig, our new Army friend. He was a gangly, sandy-haired boy from Dallas, on temporary duty (TDY) from Ft. Polk before he was scheduled to deploy to Vietnam. He ended up having to read the whole book, keeping up as we went and adding whatever he thought was key that may have been missed during our studies.

"Do you all think the Marine will get old Bart?" I put up the question to see what response would come from these guys who were already put off by his superior attitude.

Joe was first to take the bait, "They deserve each other."

Barry, Mr. Easygoing, said, "Why not? He'll be a good handler."

"Yeah, he better watch out for Bart. Bart has a record of putting away handlers." This came from Matt.

Joe took a drag on his cigarette, then dropped the butt into Vinny's beer. "I'd like to suggest a wager. Look guys, I'll put down ten dollars that Mr. Marine will not make the cut with Bart. I think Bart will be more than he can handle. I also will wager that he will be bitten by Mr. Bart."

Vinny shot Joe a bird, but Joe appeased him with another beer.

"No way," I said. "Not sure I want to take that bet."

"What kind of bet are we talking about here?" said Vinny. "I say either he is bitten and bandaged up and returns to work, or he is bitten by Bart and the bite is serious enough that it ends his dog handling career for good. I propose a side bet—does the bite end his career in the Marine Corps?

"That's damn good, Vinny, count me in," Barry said. He stuffed more potato chips into his mouth.

Next thing I knew, I was pulling ten dollars from my wallet and so were the rest of the men in the room. I still wanted to know how we'd know if we won, and how much the take, or pay-out, amounted to.

"Look guys, this has to stay between ourselves," said Joe. "Eighty dollars is a lot of money to bet against the Marine. We don't want to piss off a Marine, do we? I'm just looking for a good laugh at his expense."

Several of us walked over to the kennels and found most of the dogs resting, not even interested in making the effort to bark. The sudden quietness added something, and we whispered so as not to scare off the spirit of whatever it was going through the kennels. Even Black Bart was speechless but he never took his eyes off us as we moved past him.

"I wonder how long Bart has been on his leash this time around," said Joe. "You think he might be a good boy long enough to get out of his birdhouse and take a little run before he takes out the Marine?"

"What's the matter Joe, getting nervous?" Vinny asked.

"No, but I've been thinking, we should add what day and time Black Bart will take him out. How many days before it's his last?" Joe said. "It would make it interesting, don't ya think? Winner will guess the day of the Marine's demise."

We agreed.

Chapter Two

Duke of Earl

Lackland Air Force Base, San Antonio, Texas – October 1968

US Headline News

Led Zeppelin Makes First Live Performance at Surrey University, England.

Former US First Lady Jacqueline Kennedy Marries Greek Shipping Tycoon Aristotle Onassis on the Greek island of Skorpios.

By Thursday, we were sick of the Marine and his superior attitude and wanted injury to overtake him. He had made adversaries of the whole damn class back on the second day of Sentry Dog Training School. He wouldn't even join us for our study group, which worked out to be useful.

"You know, Vinny, not since this Marine arrived on base have I seen him introduce himself or even shake hands with any of us," I said.

"Come to think of it now, you're right," said Vinny. "This is so strange."

Barry said, "Well, it's not like we haven't gone out of our way to be friendly."

"To hell with him," said Joe.

Staff Sergeant Anderson began our class Thursday with an announcement. "Tomorrow you'll be issued your equipment and your dogs. Your job will be to get in on the dog, put a muzzle on your dog, get your dog out to the exercise area, run your dog until you are about to drop, and then run him some more."

"Excuse me, Sergeant," said a voice from somewhere in the back of the room. It was Craig.

"Yes, Private."

"Sir, that's a lot of running. Can't we maybe just play more and run less?"

"Listen, all of you, the purpose of the running is to get the animal so tired that when you remove his muzzle from his face he'll be too tired to attack your face."

"This works most of the time, so be careful," said Staff Sergeant Adamson. "Most of you will not have the energy to tire out a dog, but run him for as long as possible."

"You know, Vinny, I agree with you," said Barry. "Could these men, our beloved instructors, have a sadistic streak in them?"

"Do you think these guys are taking bets to see who gets bitten?" Joe said. "Or bets on when the Marine will meet his fate with Bart? They know these dogs and who the biters are."

"It's blood they want, and they think it's funny. 'Remember men, you're not a dog handler until you are bitten.' What a line of crap to lay on us," Vinny said. "I, for one, refuse to be bitten. It's not going to happen."

Joe said, "I agree, Vinny, and I believe there is another bet we should consider. Will Vinny be bitten before we complete K-9 school?"

Craig, our Army friend, asked an instructor how they decided which handler got which dog.

Sergeant Anderson took the question. "Mostly it has to do with size, voice and what we call body diction of both Airman and dog. Say an Airman weighs in at 150 pounds—a dog weighing around 50 to 60 pounds would become part of the Airman's profile. Heavier handlers can handle larger dogs. Temperament and over-aggressive dogs make placement challenging, we do the best we can."

Sergeant Adamson said, "The quality of a man's voice is also an important determiner in how well a handler gives commands and how well he can control a dog with voice only. That's why strong, single-syllable commands are used."

Another student asked Sergeant Anderson which of the dogs was the worst behaved of all the dogs in the kennels. We all knew which one it was.

"Bart has been at the school for more than nine months and had as many as four handlers. All were bitten at least once and one Army man more than once. All were reassigned to different K-9s. Bart is still looking for a handler."

The Marine's hand shot up into the air.

"Sergeant, I would like to handle Bart," said the Marine.

"You may get your chance, but as of right now, we still have to go over our evaluations tonight. We'll let you know tomorrow morning," said the Sergeant.

"Yes sir. Thank you, Sergeant. I can handle that dog and I want that dog."

"I hear you, Marine, but Bart may be too much dog for you or anybody in this class," said the Sergeant.

"Sir, there isn't a dog in the kennel I cannot handle. Again, I request to handle Bart."

Damn. He was nothing if not persistent. We were in pain trying to keep from laughing.

"Have patience, son. You might bite off more than you can handle, excuse the pun."

"Thank you, Sergeant. You won't regret it Sergeant," the Marine said. He returned to attention in his seat, but now he was sporting a big grin from ear to ear.

We broke for lunch. A bunch of us headed for the mess hall. The Marine sat with others for a change and was the center of attention. His stuffy, better-than-you-attitude had turned most of us against him. But the thing is, we really wanted to like him because he was a Marine and we respected him. We wanted to talk with him.

Vinny, Joe, and I sat at a table within easy earshot of him and took in all the conversation going on. For the most part no one was giving him any slack.

"Hey Marine, you better watch out with ole Bart. He doesn't much like people, especially Marines," said someone at his table.

"Watch your mouth," the Marine shot back.

My Very Own Puppy Dog

One of the nice things about hanging with Joe and Vinny and our group was being able to talk about the dog of our dreams. Our thinking was to get a good working dog that would keep us alive and get us back home in one piece. So what if the Marine had the meanest dog in the pack? We spent time during our evening study hall discussing the pros and cons of working a dog like Bart.

We were talking about our dogs attacking us while under attack by Vietcong. What if my dog turns on me? And we're under attack by the Vietcong? We may find ourselves in a world of hurt, in the blink of an eye, a most deadly situation.

Eight weeks of school did not leave much time to deal with a psychotic dog. We all agreed that our Marine friend could be making a serious mistake; it could get him killed and take other Marines with him.

"I don't trust Bart as far as I can throw him," said Barry.

"I hear you, Barry," I said. "I don't think Bart likes humans because he keeps biting his handlers."

Just because a dog is mean doesn't make him stupid. It's easy enough to make a smart dog into a mean, vicious dog at a command; command being the key word. Bart didn't take to being controlled and less to being commanded.

I did not want to handle Bart. No way.

It was too bad the Marine didn't have another Marine around to talk to—he wouldn't listen to any of us Air Force pukes. We tried to tell him how important the nose and ears of a German Shepherd were compared to how he looked and acted, but he just sneered at us with Marine contempt. He wanted the image that the most ruthless, meanest dog in the jungle would give him.

"Can you relax sitting at attention?" Craig teased the Marine in the classroom one afternoon.

He looked around with a strange smile. "It's a Marine thing and you're not expected to know or understand, so leave it alone," he said. It wasn't funny at all but the classroom burst out in laughter.

Attitude or not, I knew enough to keep my thoughts to myself. *Are you shittin' me? What kind of crap is this?*

At least once a day that first week of Sentry Dog Training School I made the short walk back and forth from classroom to the kennels.

Which one of these beautiful animals will be mine? I wondered.

A few dogs would still bark and act crazy. Bart paced nonstop back and forth, around and around. As I stood and watched this beautiful animal one afternoon, he didn't bark at me for some reason, nor would he take his black, pitted eyes off me, either. I'd rather he had snarled and howled, giving me an idea of his head, (thinking).

The point being, never turn your back on one these animals, not until you became close friends.

Thick, heavy chains made of steel were used to secure the dogs to their birdhouses, but chains couldn't be trusted. Even steel chains could break at any time and if an Airman was not paying attention, a dog like Bart would be all over him in a flash. I moved on. It wasn't worth taking chances with him.

For the first time in two months, my thoughts didn't center on Lillian as much, but that made me uncomfortable. *How can I forget someone so quickly? I'm not forgetting her, it's just something else is trying to take her place and that scares me.*

It was about ten hundred hours on Friday morning and almost time for us to be assigned a brand new puppy dog.

First, Sergeant Lotes gave us a briefing about our gear, which included a 30-foot leash, a 5-foot leash, choke chain, muzzle, collar, comb and brush. We were also issued a web belt with an attachment ring for holding all the gear and two water canteens. No guns.

"You must care for the leather muzzle, two leashes and collar by applying oil regularly," said Lotes. "It makes the leather pliable and comfortable for the dog's face and our use, too. The more you work the oil into the leather, the softer and more pliable it becomes."

The kennels were in total chaos as new handlers were introduced to their dogs and started the process of trying to get close enough to even try to put the muzzle, then leash and choke chain, on these agitated dogs. The first thing we had to do, however, was get inside the circle. The circle was the perimeter around the birdhouse that the dog could move around in, *the Circle of Death,* I thought.

Sergeant Anderson was escorting me to my assigned dog when he came to a dead stop in front of Bart's birdhouse.

Next to Bart stood a frozen, pale Marine. Sergeant Anderson's dark eyes fixed on a muzzle that lay on the ground between the dog and the Marine.

"Moore, stay here and don't move," said Sergeant Anderson.

Bart was by the Marine's side looking up and growling, showing teeth and ready to pounce.

"Sergeant Adamson, need your assistance over here, quickly please," said Anderson.

"Listen up, Marine," said a livid Sergeant Anderson, "On the count of three, Sergeant Adamson will distract Bart long enough for you to run like hell. You got that, Marine?"

All the Marine could do was nod his head up and down in the affirmative.

"Sergeant Adamson, on my count are you ready? One... two... THREE," said Anderson.

Adamson started yelling, his wiry arms flailing while he jumped and charged like a crazy man. It worked. Bart charged the Sergeant and the Marine ran for his life. Just as the Marine cleared the circle, the steel reinforced chain cracked, Bart's jaws snapped at the fleeing handler.

Both Sergeants dressed him down with an Air Force chewing out. "Marine, what was the number one instruction you received for entering and controlling a dog on his turf for the first time?" asked Sergeant Adamson.

Before the Marine could answer, Sergeant Adamson barked at him again. "Put the damn muzzle on, first thing. Got it, Private? And another thing, you've made it that much tougher on yourself, Bart has your number now."

It was over. Sergeant Anderson and I moved on down the line of birdhouses, stopping in front of a growling, snapping, red and black German Shepherd. This was my dog.

His name was Duke and his service number was X010. He stood almost waist high and his coat was black on reddish-blonde. He didn't have that long-body-on-short-legs look that most Shepherds have. Instead he was tall, compact, and lean, with big ears that stood erect, and soft brown eyes. He was the most beautiful animal I had ever seen.

Damn it, he's mine. The Air Force has assigned him to me and now he's my responsibility. This is scary and humbling all at the same time. What now? Get him out and go for a run and play. "Would he really turn on me?" I whispered to myself.

"Damn right, he will," said Sergeant Anderson. He was reading my mind.

Sergeant Anderson left me staring down at Duke and the first thing that came to mind was the song, "Duke of Earl." I started singing, "Duke, Duke, Duke, Duke of Earl. Duke, Duke, Duke of Earl." I couldn't get the music out of my head.

Duke the Dog stopped running around and just stood there looking at me. I stopped singing and he took off running. All I could do was just stand there and stare at this dog as he barked and growled and ran around in circles like the rest of the dogs in the kennel.

"Come on, Duke. Settle down, boy," was all I could think of to say to him. I hummed to myself and took a good look at the dog the Air Force had entrusted to me. I was a working dog handler earning my pay. Duke was something else, and I was ready to go for a run.

If only Duke will cut me some slack and get on with the business of training me.

"Duke, you've done this all before. You are the trained one here, so cut the showboating and let me take you for a romp in the field," I said.

Duke continued to run around his birdhouse, stopping only when I got a little too close to the Circle of Death. He would stop and charge at me, over and over, for what seemed forever. He tired me out just watching him. I sat myself on the ground about two feet from the end of Duke's chain and started the slow process of talking him out of mauling me.

Thank God for strong chains! I hope that all the chains in the kennel hold their beasts. Will Duke bite me today? I wonder how many of my classmates will become initiated K-9 handlers today.

I said, "Duke, you are one beautiful dog, and I'm your handler now. I will take you out this very morning before lunch and that is that. Put that in your pipe and smoke it."

Duke was no Bart. He would stop barking and scratch some part of his body, paying more attention to what was going on in the kennel than to me. But just as soon as I crossed the circle into his space, here he came charging, barking and growling, showing off a big set of teeth. I stepped back and watched him, all the while talking to him in a low, soothing voice. Duke was barking less now and seemed calmer, somewhat.

Duke's behavior was a product of his environment. All the dogs in the kennel were going nuts, for the place was a madhouse of growling, barking dogs running around and around as they circled their dog houses. Just one dog used to drive me nuts barking; now there were well over a hundred and it didn't bother me at all. I knew that if I could get Duke out and away from the kennel area we could both relax. My fatigues were drenched with sweat—and it was still cold that morning.

He sensed and smelled my fear. Without thinking about it, I crouched down and started singing "Duke of Earl" again. Duke stopped what he was doing and stared at me. I moved up to his boundary line, crouched down still singing, and here he came running, but he didn't bark this time. A good sign, I hoped.

He ran until the chain was pulled tight and stopped in front of me about two steps. I kept singing and moved a step closer to him. He was in my face, and I was looking past teeth and tongue and halfway to his stomach. He kept barking and pulling on his chain, and I prayed the metal would hold. When he barked at me, Duke would look around as if to see what he might be missing. I took this to mean Duke wasn't such a bad-ass after all, just showboating.

Looking around the kennel area, I saw that the other handlers were having as much success as I was, so I began to relax. Sergeant Anderson walked by and handed me some treats, telling me not to say a word. Screw the rules.

"Airman, are you singing to Duke?"

"Yes, Sergeant. Not sure if it will do any good," I said.

"It won't hurt, Airman. Carry on."

Sergeant disappeared and once again I was left alone with Duke. After I spent an hour talking in a quiet, soothing voice, he relaxed. If I attempted to step inside the perimeter of his space, Duke came charging at me, snapping, growling and barking. This had gone on for several hours and I was tired, thirsty and frustrated. Duke was messing with my mind.

This is kind of petrifying and funny all in the same breath. Will this dog really attack me? Is it so bad getting bitten and bleeding all over the place? Yes, it is.

Duke had trained for sixteen weeks at the least and knew what was expected of him.

Me?

Not a clue. I knew that Duke was trained to attack humans, take a man to the ground, and secure him there.

Do they want to kill and eat or is it just a job? Once they taste blood, does the job description change?

Back home, if we saw growling and snapping dogs, especially if they were running in a pack, we grabbed a gun and used it if necessary. Parents worried about children playing all over the neighborhood and a dog or a bunch of dogs biting one. We also knew never to run from a pack of dogs.

Looking around the kennel, I remembered the wolves and how they hunted in packs. Anything could happen. *I have one job to do, get that blasted dog muzzled and on a leash and out the door.*

Duke liked the treats because they were tasty and I liked the treats because they calmed him down. I thought that maybe, if I walked into his space and caught him off guard while he enjoyed a treat, then I could muzzle him. That seemed doable.

Let's see, so far, I have sung to him, talked calmly and let him know I am a friend, and I gave him all the treats except one. No muzzle on him yet. Damn.

The sun was tilted low in the sky and still, no Duke on a muzzle. If unsuccessful, I'd be back Saturday with the rest of the class members that couldn't get their dogs out of the kennel on Friday. Duke was about to cut into my social life. Feeling discouraged and having the need to just hug him, I went on with the plan.

Duke was no longer barking but he never took his eyes off me, either. Using that eye contact, I held out the last treat in the palm of my hand and walked into the Circle of Death.

At first Duke was confused but moved in my direction on instinct alone, never taking his eyes off his treat. I had the muzzle ready as Duke accepted the treat, his mouth snapping it up.

Here's my opportunity. Duke can't chew and bite at the same time, I reasoned, but the muzzle did not cover his snout as planned. The leather was so new and stiff that I failed to get it over the ears to fasten the muzzle. I was so nervous that I dropped it.

Oh, hell! Lucky me.

Duke continued chewing, paying me no mind, so I snatched up the muzzle, tried again, and this time was successful. Next came the choke chain that slipped over his head, and then I snapped the sixty-inch leash to his chain.

By this time, Duke had finished eating his treat, seated himself on his hind-quarters, and stared up at me. My heart raced and I enjoyed my first victory over myself as much as this K-9. I managed to remove the large brush that was connected to my web belt and began to brush Duke.

He just stared at me and I gazed into brown eyes that seemed to say, "These are friendly eyes." It was then that I gave him his first hug and he leaned into me as I brushed him.

Duke's reddish and dark brown coat was broken up with spots of black and some white on the chest. His tail was all red, thick, and long, and stuck straight up in the air. His face had some blonde and white under the eyes, as well as red. His ears were perfect and stood up the way a good German Shepherd's ears ought to.

"Duke, you are the best-looking dog in the kennels." I hadn't realized I was still singing to him. "Duke, Duke, Duke, Duke of Earl."

"Excellent job, Moore. Now, get him out to the exercise field and don't forget to change out the choke chain for the collar. If the dog gives you a hard time, then leave the choke chain on until he gets used to you," said Sergeant Anderson.

How long has he been watching?

"Yes sir, Sergeant," I said. I'm sure I had the smuggest-looking expression on my face as I passed guys still trying to get their dogs to cooperate. Damn, it felt good. I started singing "Duke of Earl" again as we moved out, this time as a team.

Duke relaxed and fell in line next to my left leg, where he was supposed to be. Indeed, Duke had been trained and was about to start training me.

"Way to go, Moore! Good job." I heard from several of the other handlers as I walked Duke out of the kennels and over to the open field. I was on cloud nine.

"Duke, you're a damn fine dog and thank you very much for not biting me," I said. "I'll just have to learn to be a dog handler the hard way, so have patience with me, Duke."

Duke was coming around and I could tell he was happy to be away from his birdhouse. I left his muzzle on, not ready to take any chances. We walked and ran for about an hour. He seemed to be playful enough so I decided to remove the muzzle. We walked to the far side of the field and with a lot of praising and loving him up, I dropped to one knee and removed the muzzle.

Whew! No attack! So far so good.

I had snuck a pork chop from the mess hall after lunch and stuffed it into a pocket. Duke could smell it on me and was delighted when I pulled it out and allowed him to take it out of my hand. He got too insistent and I yelled, "No!"

It worked, and he gently took the chop in his mouth and sat down to enjoy his snack. I sat next to him to watch him eat, then gave him some water when he finished and let him relax for a little bit. Then we made one lap and then turned for the kennels.

Word came down to call it a day. Joe walked over with his dog Prince and we sat down for a chat while we groomed our dogs. "I haven't seen Vinny on the field yet," Joe said. He lit a cigarette.

"Vinny has not gotten his dog to cooperate. He is not a happy camper, he has to report for duty tomorrow." I said with a sly grin on my face.

"There are several other handlers pulling duty on Saturday so he won't be alone," said Joe.

The twilight was fast disappearing and the kennels would have to be cleaned up. I'd bought Duke off with treats and a pork chop, but the final test for the day would be getting him back to his birdhouse and me safe outside the reach of the Circle.

He may not be ready for me to put him up for the night, not yet. He may decide to attack.

After our final run, I gave Duke a fifteen-minute cool down period before I put the muzzle back on. The muzzle went back on the dog any time the dog team went into a public area, like the kennels. I put Duke in the sit position, unbuckled his muzzle where it was attached to my web belt, placed it over his snout and ears, and buckled it around his neck. There was no resistance.

We walked into the kennel area and Duke appeared to be prancing, his tail standing tall. *I guess being proud of his handler makes as much sense as me being proud of him. And I am happy to have him, looking forward to seeing what he can do and what he will be teaching me.*

I approached his doghouse, walking into the Circle of Duke's space as if it were not there, right up to his birdhouse. Without hesitation, I put him in the sit position and removed his muzzle. I put on his collar, removed his choke chain, and reattached him to the birdhouse chain. I brushed and groomed him for a few minutes, then got him to lie down as I thought about how to make my escape.

I was playing with him and noticed that his water bucket needed refilling. I got up, walked away and grabbed the bucket, and headed for the water hose. I knew Duke would enjoy some cool, fresh water.

While returning after refilling the water bucket I looked down the long line of birdhouses and saw Duke eyeballing me; it was then that it sunk in what I had done without realizing it. I had turned my back to Duke and walked out of his area without thinking about the danger.

Lucky there are no instructors around. Well, it's too late now. Just put the bucket down and leave.

I sat the bucket down for Duke, who was waiting for a drink; I petted him a few times, then stepped out of the Circle and left. When I looked back, Duke was still drinking, not paying me any attention. That concluded my first day and my first ever attempt removing a muzzle and putting Duke back in his kennel.

"Duke, Duke, Duke, Duke of Earl, Duke, Duke, Duke of Earl," I sang. *My God, how I love this job.*

Peer Pressure

Saturday morning, I walked over to the kennels to check on Duke. A hundred feet from his birdhouse I yelled his name and he noticed me. By the time I reached him he was wagging his tail and glad to see me. I ignored him and instead picked up his water bucket and patted him on the head as I walked by him. Upon my return with fresh water, Duke was jumping and barking and his tail was going 90 miles a minute. Oh man, was I glad to see that reaction.

In Vietnam, anywhere actually, a handler must have the love of his dog. It was as important as c-rations or ammo for weapons. If a sentry dog sensed love, there was nothing that K-9 wouldn't do for his handler, his friend.

I put Duke on his leash and muzzle and walked him over to the obstacle course to see how he would do. He completed it without too much prompting, but once he got to running I struggled to keep up.

"Duke, my boy, you done good. Good boy," I said. I kept on petting and talking to him.

He would jump up and run like a bat outta hell, back and forth, and at the command, "Duke, come," he'd approach me full of excitement and start barking and wagging his tail. He licked me instead of biting me and that was a big win. It didn't mean I could let my guard down, but I sensed the bonding and Duke did, too.

When we were done, we went back to the birdhouses. I cleaned up Duke's area and again gave him fresh water. I was determined to put the idea in his mind that I was now the one who took care of him, working on trust.

"Duke, you are one outstanding dog," I said as I petted him. "Thanks for not biting me, today. I'll see you later, alligator."

Even after his attempts on Saturday, Vinny was unable to get his maniac muzzled and on the leash. At the club that night, Joe said, "Hey, Vinny, when you gonna' man up and show that dog who's boss?"

He took the ribbing, smiling, he had no choice.

The next day being Sunday and not an official work day, we didn't have to go in, but Vinny decided he would go to work. He was mortified thinking on Monday he'd better have his dog under control, or face the entire class ashamed and feeling like a failure.

On the exercise field there were several teams with muzzles on giving the handlers fits. Russell Puls removed the muzzle, and his crazed beast went straight for the first available flesh, his hands.

"OK, Airman, shit happens, you're lucky you didn't get nailed." Sergeant Anderson had been observing him at a distance but began moving closer in case he was needed.

Airman Puls looked like he wanted to crawl under a birdhouse.

"Put the muzzle on and run him. Take Lucky over to the obstacle course, but I wouldn't remove the muzzle. Use it as punishment for today and try again tomorrow."

I don't want to be the student handler whose dog bites an instructor. Lord, please let me get through this class without looking like a dumbass.

We made jokes about the instructors getting bitten. "How many times does one of our instructors have to be bitten before being promoted?" We kept it to ourselves.

The second week of school was a time to get acquainted, exercise with our dogs, and always obedience, obedience, and more obedience. Several of the handlers already sported wounds but nothing serious. On Monday morning, Duke and I ran around the two football-field-sized training areas where we would train for the next three weeks.

Some of the teams have muzzles on, still. The Marine and Bart are doing okay, but Bart has his muzzle on. Glad to see the Marine is using his head.

After lunch, Vinny and King were at last out of the kennels and running with muzzle on.

Barry groomed an unmuzzled Sacha and Joe took a smoke break under a tree with Prince, no muzzle. From their vantage point, they heckled Vinny. "What's the matter, Vinny? Your doggie still wants to eat you?"

I stopped to catch my breath and brushed Duke's coat, continuing a steady stream of loving up and petting. I'd have given him a treat if I had some, but we were not allowed to.

Sergeant Lotes had warned us the first day we met our dogs. "As much as a dog loves a treat, getting that dog to perform with the aid of food is a mistake. In the bush, where a dog has alerted, an Airman may come under fire as he investigates. That is no time to praise and offer up a treat for doing what the dog was trained and expected to perform."

On a Saturday or Sunday morning I sometimes would take over some sausages from breakfast. No treats for tricks, no confusion, and the dog would better retain his training when I needed him most, like coming under enemy fire in the middle of the night.

Out of sixty-two dog teams, I saw only one handler having trouble with an aggressive dog that day. The handler had the dog face down on the ground, pinning him. The instructor stood by and said nothing because man had to win over dog if he was going to be of any use to the Air Force.

By the end of the day, Monday, all muzzles were off.

Tuesday morning at ten hundred hours the roach coach arrived for break time. As the line formed up, Joe with his dog Prince and

Vinny with his dog King were in front of me. Both dogs seemed calm and easy to work with and neither handler was having any control problems.

We stood around drinking soda and noticed Bart trying to get something to eat from his Marine handler. Bart leapt up and snapped his jaws around both the chips and the Marine's hand. He wasn't turning lose. As the dog came down with the chips, he dragged the Marine to his knees.

Several of the instructors ran towards the Marine. His hand had been severely injured and was bleeding. Sergeant Adamson barked an order. "Muzzle that dog, Marine!"

He moaned. "I can't."

The bleeding was bad. He couldn't unhook the muzzle from his web belt.

Sergeant Anderson grabbed a muzzle from another handler and moved from behind Bart, who was still busy eating chips. He didn't see the Sergeant coming until he had pinned Bart to the ground. Adamson secured the muzzle on Bart and walked him away from the Marine.

The other instructors lifted the Marine to his feet and began wrapping his hand in a bandage. His wound gushed blood until his uniform was soaked. A rescue unit turned into the field and pulled up next to the injured Marine. The instructors were saying Bart had torn several of the Marine's ligaments and tendons. He'd be lucky if he didn't lose the use of his hand.

"Did Sergeant say the Marine won't be back?" I asked.

"That's what I heard," said Joe.

"I bet it hurt really bad. Dumbass Marine," I said. We looked at each other and almost burst out laughing.

"As you were," said Sergeant Anderson. "He has a serious injury. That's it, get back to work. I'll tell you more about him as we hear it."

Bart had taken out another handler.

After we finished out the day, we hung around the kennels longer than usual to see if we'd hear anything about the Marine and Bart. There were a lot of reasons for not liking him, but no one wanted his dog-handling career to end like that.

Joe said, "Damn, how do we figure who's going to win the pot?"

"I don't know, Joe," said Barry. He paused. "Something's not right about this."

Vinny suggested forgetting the bet. "Damn, did you see his hand? It was mauled. He may never handle a dog again. And what if his trigger finger doesn't work? No, I'm for returning the money and forgetting this shit."

"Well, the asshole didn't deserve what happened," I said, "and it's kind of sickening, but the thought of his Marine career being over is what I'm concerned about. It would kill him. This whole thing is a bummer now, and I'm with you, Vinny,"

"Hey, Joe," said Vinny, "if you haven't already spent the money, how about returning it so we can get away from this incident, just too freaky."

"Sure, OK."

The following week, class convened to hear Sergeant Anderson announce the next stage of training.

"During agitation training, you men must have complete control of your dogs at all times. We're in the field now. This is where the rubber meets the road, and people can get hurt."

"To start, all agitation training exercises will begin from a changeover, choke chain to working collars," said Sergeant Adamson.

Our first exercise was line aggression. No muzzles added to the excitement. Adamson explained the procedure. "The agitator will use a protective burlap sleeve which the dog can bite into, and will aggressively rouse the dog to the point of wanting to bite. The handler never agitates his own dog, and we never agitate from a vehicle. We use a leafy switch and burlap bag to provoke the dog without actually striking the dog."

Sergeant Anderson demonstrated with burlap rag or switch in hand, moving down the line towards and away from the dogs. We all hoped some handler didn't lose control of the leash.

For the next exercise, Sergeant Anderson, the agitator, was in the center of the circle with a burlap bag and his trusted switch. The circle was larger than for obedience training.

"The purpose of this exercise is to give the dogs the opportunity to go after their antagonist," said Sergeant Gildart. "Dogs love the chase, but they also love a good take down or capture or to sink their teeth into something. Burlap fine, flesh better."

Anderson started to move towards the animals. As he passed by each dog and handler, the Sergeant screamed, waved his hands with the burlap bag and switch, charged the dogs and backed away. This was driving the dogs mad. It took all our strength to hold on to the leashes. The hair stood up on my neck, my heart raced, the noise was ear shattering.

Could this be better than sex?

"Men, slowly now," the Sergeant said, "slowly move toward me, not too fast, keep your distance."

The manual said not to strike the dog, but we learned from our instructors to break the rule in some cases. If a bored dog scratched himself or seemed to be longing to take off after a squirrel, Sergeant Anderson would take out the switch to tap them across the nose. The branches were soft enough that they couldn't hurt the dog, but being hit in the face over and over would make him meaner than a rattlesnake getting rocks thrown at him.

"The belt line using a leather collar across the backside will get the attention of any breaking dog quickly," Anderson said.

Dogs breaking and going after another dog was a serious no-no, and manual or no manual, we dealt with it. Even if the dog could not see the handler or, for example, a suspect stopped running and put his hands up, the dog must respond to commands.

Staff Sergeant Anderson got our attention. "Look men, keeping a high level of training is imperative. We nip bad behavior in the bud."

"Well, the belt line sure as heck did that, Sergeant," I said. "By quitting time, we could have walked fifty yards and I don't think a single dog would have broken."

"Belt line. That's a good name for it," said Barry.

"Damn straight," said Vinny, "Sure handled that Navy guy's mutt.

"I have no patience for a dog who has cycled three times through this program so far and goes after another dog," said Sergeant Adamson. "Makes me look bad."

After a few rounds of the belt line it sunk in. Most of the dogs did not break and go for the bait dog, including Duke. I was proud of him.

Friday after our lunch break, all thirty-one dog teams in our class were moved to the far side of the field. There was a tree grove with several huge oak trees and we were directed to get in a circle under the trees. Thirty-one handlers with dogs makes for an enormous circle.

"So, tell us, how do you like your first week working with your own dog, as a sentry dog handler?" asked Sergeant Anderson.

A big rebel-like yell went up and the dogs started barking. It took several seconds before order was restored so the instructor could continue.

When we settled down, Sergeant Adamson went to the center of the circle and began speaking. "Men, as your teachers we have an obligation to give you as much knowledge and know how as we can. We will be introducing a technique today that some of you may find a little bizarre, but we think it is necessary to put this idea in your bag of tricks for use in Vietnam."

This is different. I wonder what it could be?

"Sentry dog handlers do not and cannot depend on giving treats in order to get our dogs to follow instructions," said Adamson. "Is the handler going to offer a treat to a dog who is engaged in drawing human blood from an enemy?"

Several of us shook our heads "no" in agreement with the premise.

"I want each of you to get down on one knee next to your dog, and with your left hand take your dog's pecker and masturbate your dog."

Most of the class froze in place, too embarrassed to act. A few started to go down on one knee, looking around to see what the others were doing.

"Are you shitting me, Sergeant?" said Craig. "What will my mother say if she ever finds out about this?"

The class rumbled and stalled, but Adamson said, "Listen up, men. Your lives depend on these dogs following instructions in the field."

Sergeant Anderson marched to the center of the circle. "There are only two reasons for not following our instruction," he said. "Number one, you object for religious reasons. That's okay. There will be no questions asked."

I think there is something in the Bible about this, as in bestiality. And what would my parents think? That's easy, they'll never hear about it.

"Reason number two, your dog is a bitch. Now drop to your knees and jerk your dog off."

Several trainees laughed nervously, attempting surreptitiously to observe what their classmates were doing. "Come on, we don't have all day," Sergeant Anderson barked.

Oh my God, this is a Twilight Zone experience that Rod Serling wouldn't touch. Is the way I do it to myself the right way to do it to a dog? Playboy hasn't covered this part of the sexual revolution.

I looked around and saw most of the handlers were busy bringing pleasure to their dogs. Sergeant Anderson was standing behind me. "Get on with it, Moore. You're wasting time."

Duke was looking at me with those big brown eyes as if to say, "Well, get to it."

"Airman, it's no big deal. Just grab ahold of that furry part and jerk him off. Believe me, he will be a better dog for it."

"Yes, Sergeant." By now, most everyone else was doing it. Peer pressure won out in the end.

It didn't take long for Duke to finish. I glanced up to take in the whole circle of happy dogs and began to laugh. Only Russell still seemed confused and had not started yet.

"Don't worry, your dog is female. Watch and remember," said Lotes.

Russell's white pimples stood out against a beet-red face. To a man, everybody did as instructed except Russell, who had the bitch.

By now, 24,000 young men have died in this war. I don't know what our instructors could ask of us that we won't do. This is our ticket to get back in one piece.

"Hey Sergeant, is my dog's pecker broken?" I said. I felt stupid for asking, but I was worried because the furry part had not gone back to cover Duke's penis yet.

Sergeant took a closer look. "Airman, don't worry about that. It will go back to normal. Just leave it alone."

For the rest of the afternoon we were given free rein to work with our dog as we chose.

"Hey Joe, did you catch the guys by the fence? Man, they are giving their dogs a good working over," I said.

"Yes, I did," Joe said. "Poor dogs."

"I think I'll be passing on this," I said. "If my date finds out what I've been doing in school, there will be no more dancing, hugging, or kisses."

"Hey, I can hear it now." Joe said. "You did what?"

Chapter Three

Ashes to Ashes

Detection and Aggression Training, Medina, Texas – November 1968

It took two trips to move thirty-one dog teams to Medina, just a few miles from Lackland. Medina was where we would have the Detection and Aggression phase of our training. It would be our home for the next four weeks until graduation day. We would become hunters of men, a serious, deadly force.

Sergeant Anderson explained the first order of business. "First thing we're going to do is clear the area of snakes," he said. "Rattlesnakes."

We Airmen thought it was a joke, but the Sergeant's brown eyes were serious.

"This is another part of your training to prepare you for snakes in Vietnam, like cobras," said Sergeant Gildart. His tightly pursed lips did not even hint at a smile.

Shit balls! I hadn't even considered snakes in Vietnam.

I could hear whispers of, "Oh, damn," among our group. Out of the corner of my eye, I saw Joe running a nervous hand through his thick, wavy, black hair.

"During construction, nests of rattlesnakes were unearthed by the bulldozers when they built the kennels and cleared land for the training areas."

So now I have to be a snake hunter. What the fuck? Over.

"Don't worry if you see one; in fact, you probably won't find any but we have to warn you men to be on the lookout," Sergeant Anderson explained. "Just back away and call for one of us."

"I hate snakes." Joe said. "When I was in the sixth grade I was sitting on the back porch when a big black snake crawled across my foot and stopped." He paused a moment. "I'd still be staring at it if my dog hadn't started barking and chased it underneath the house. I was scared for weeks after, scared I was going to meet up with him on my way to school."

I grinned. "Vince, you got a favorite story?"

Vinny fussed with his collar a little. "No, but I hate 'em as much as anybody does, they're just so damn ugly."

We heard Sergeant Anderson approaching behind us, or rather, we heard the "scratch-scratch" of his crisply starched uniform. "You boys like fried chicken, don't you?" he asked. "Then you'll like fried rattlesnake."

You're kidding me, now I've got to eat freaking snake? Yeah, get me ready to face a variety of snakes in Vietnam deadlier than anything in Florida. Shit. I don't want to do this anymore. Cook school, bunkers, towers, even riding around in a track (APC) all night is looking better all the time. I'm screwed.

Monday morning of our seventh week we gathered in the classroom for a briefing of the schedule for week seven.

Sergeant Anderson brought the room to order. "At ease, men." Sergeant waited as the room grew quiet.

"I've got a lot to cover in the next hour or so. I'm assuming all of you have read your manuals, right?"

We nodded.

"Ok, week seven is all about scouting and patrolling and learning to read a dog's alert. Not just your own dog's alert but watch each dog to see how he alerts. Especially watch to see if the handlers pick up their dogs' alerts."

For a moment, we stole nervous glances at one another, while Sergeant Anderson continued pacing up and down our straight rows of tables.

"If a handler doesn't know his dog's alert, what good is he? He is good for nothing and of no use as a Sentry Dog Handler. He will be out of the program, shipped to Vietnam, and most likely will be in a bunker or a tower where Charlie will know right where he is. Not a good thing, right men?"

No question about it. I didn't want to be in a bunker or a tower.

"Men, this is the week that will kill you if you are not on your toes and paying close attention. Stay focused." Holding up the training manual, he continued. "Okay, now, let's talk about scouting, patrolling and gunfire training, straight from the manual."

Sergeant Anderson is scaring the hell out of me. None of us has ever seen an alert, not even our own dog's. Little greasy shit-balls, what next?

"In Vietnam, most of the time you will be securing a perimeter or fence line, just you and your dog on one side and the enemy on the other, some of the times."

Anderson paused, his body tensed, as if he were out on patrol back in Vietnam.

"The enemy is an expert at getting on your side of the fence. That is why the Air Force uses K-9s and we have a good record in Vietnam."

Shit! It's about time these men start telling us what it's like over there. Hells-bells, what should we expect? Come on, now!

Adamson placed his flat hand at chest height to illustrate. "Your enemy stands about four feet tall," he said.

"He weighs a hundred pounds, and will come through any fence line undetected wearing nothing but underwear. He will spread axle grease on every inch of his body thinking that will cover up his body scent from the dogs."

The classroom was silent with tension. Any talk of VC or Vietnam made my back stiffen.

Sergeant Anderson stepped in again. "So far it hasn't worked very well for the Vietcong, but they keep coming and they keep dying. Every once in a while, they get lucky and kill a handler and dog, most likely it will be the dog they kill. All across Vietnam there are bounties on dog teams. Killing the dog is more important than killing the handler. I wonder why," he said, raising an eyebrow. "Learn your dog's alert, men, and come home alive. It's that simple."

"What is the incentive for your dog during obedience training?" asked Sergeant Lotes.

"Praise and loving up your dog." The chorus came from half the men in the room.

"That's a good answer, but in the manual, we call it verbal and physical praise, correct? What is the incentive for a dog to scout or search out and detect an intruder?"

It was like a shout for joy as most of the handlers howled, "The chase!"

"Very good, men, damn good. I think most of you have read the manual," said Lotes.

On Monday afternoon, Sergeant Adamson introduced us to the suit.

"Men, meet the suit, all seventy pounds of inch-and-a-half thick burlap. It's made from the finest materials, made to fit over legs, body and arms."

Up to this point, we had been using the switch and burlap bag on our dogs and the instructor had used the sleeve. Now it would be our turn with the sleeve and to climb into the padded suit.

"Sergeant, it looks like these suits are hard to move around in," Joe said to Sergeant Adamson.

Sergeant Adamson said, "Good eyes, Airman. You're right, this burlap is almost two inches thick. Very hard to move in and you still have to maintain balance."

He walked about with stiff arms and legs to demonstrate how the suit affected movement. "Once you are knocked down, the dog knows the defenseless areas of the suit, neck up, balls and ankles. These dogs are trained to attack the suit and they love it."

Sergeant Anderson said, "Always brace yourselves against the impact of a ninety-pound dog hitting you at full speed, falling will expose all those areas that the suit would not protect."

When it came my turn, I had watched enough from the sidelines that I thought I had it under control. Nothing to it. That was until I got into the suit. It was like being stuck in an oversized strait jacket.

I couldn't move, it was so hot I couldn't breathe. Couldn't imagine being in this thing during the summer months, thanked the Lord it was winter.

Sergeant Anderson said, "Moore, I'll be right behind you with this sleeve just in case this dog goes for me instead of you."

Are you shittin' me? I was scared to death that I was going to fall the minute any dog hit me. With all the bullshit aside, the order was given and the first dog came charging at me. As he ran towards me, he leaped high at full speed.

All I saw were teeth coming towards my face. I had just enough time to put my padded arm between my face and the oncoming dog. He hit me full force with all four paws landing on my chest. I felt its teeth sink into the padding on my right arm with its jaws tightening like a vise. Through all that burlap no less. Powerful jaws.

How am I still standing?

I managed to take a step backwards, just enough to balance myself. I felt the Sergeant from behind giving me some extra support to let me know he had my back. From then on it wasn't so bad. The dog growled, snarled, and did everything but turn loose. The handler finally stepped in, gave the command "out," and the dog released.

Just like that, I had been attacked by my very first dog. It felt like being hit by a VW.

Tuesday, the instructors continued their lectures on detecting alerts.

Anderson picked up from where he left off the day before. "The wind is probably the most important and at the same time, the most erratic factor the handler has to contend with when utilizing a sentry dog to the best advantage."

He stopped pacing long enough for us to take in his words, then asked, "Why?"

In one voice, we answered. "The scent is in the air, Sir."

"Very good, men." He paused again, and his brown eyes glanced upward as he recalled his experiences in visceral detail.

"In Vietnam, you want be able to toss grass into the air in the dark to find which way the wind is blowing. Your eyes will water in a half way decent wind. Your sweaty skin will be cooled by the wind, telling you which way the wind is coming from. We've all watched tree limbs as wind passes through. When patrolling, you work from the downward side of the oncoming wind."

Vinny was taking notes, and I was glad for that. I didn't recall seeing these details in the manual.

"With no wind at night," Anderson said, "you still must clear your post. That is what quartering is all about, it helps you break down a large post and clear it of intruders. If there is no wind, then the second-best detection device is the dog's ears."

I looked over at Vinny's notes. So far, he had written, "Tools for detecting the enemy. #1 – nose, #2 – ears." Anderson was giving us all the benefit of his experience.

"Usually a dog will pick up a scent before he hears the enemy. No wind, then pay attention to the ears, the head will follow the ears. Too dark to see what the ears are doing, then gently place a hand between the ears, and notice the movement."

Not much was said about a dog's sight because it isn't a dog's best use of his senses, except at night. K-9s see in grays, day or night. During daylight, the eyes don't see as clearly but at night the dog's eyes see movement. Stay motionless anytime day or night and the dog will run right by, but if he hears a sound or picks up a scent, someone is getting nailed anyway.

"When it is pouring down rain, how do you read your dog's alert? Rain is a good time for VC to attack." Anderson stopped his pacing to recall another experience.

"Just walking the rough terrain and staying on your two feet will be overwhelming until you get used to it. You got a dog leash wrapped around one arm and hand," he motioned with his left hand as if holding a leash and with his right to simulate wielding a gun, "and a weapon in the other and the very next step ends with you falling on your ass. When it rains, it pours, then you got double trouble, and you're supposed to read an alert?"

"All right, let's break and be back in fifteen," said Anderson.

The Final Exam

Friday morning, it was time for a briefing on our eighth week, which would be devoted to field testing.

"Okay, men," said Sergeant Adamson, "Listen up for instructions for your final test."

Here was his usual dramatic pause; we called it "sergeant's-pause," and right on cue. The instructors did this when something very important was about to be said.

"Each team will search out and detect a hidden intruder. The intruder will be suited up and will try to evade by moving away from the dog team, at which time the dog will attack. Any questions?"

Barry's hairy arm and hand went up, and Adamson acknowledged him.

"Yes, Sir," said Airman Teagan, "How can this challenging exercise be conducted in such a confined area as our training grounds and with so many handlers and dogs idling around?"

"Good question, Barry," I said. "So mostly we're being tested on identifying our dog's alert?"

Sergeant Anderson took the question. "Yes, and, do you do the changeover correctly? Do you use the wind correctly while patrolling and quartering? When your dog finds the scent of the intruder, do you catch the alert? Finally, can you secure the post?"

"Good questions, shows you're paying attention," said Adamson. His blue eyes flashed approval. "You're right in your assessment. We've figured out the best way to test and it gets the job done."

It was the last drill of the day; the sun was just touching the top of the trees to the west that Friday. We looked forward to checking out the ladies at the club tonight, but something else was on Vinny's mind now, the very dangerous job at hand. It was his turn to be in the suit.

I watched from a safe distance. Vinny was in position and ready to go. Sergeant Anderson was agitating the dog with the switch in one hand and the sleeve on the other arm. As Anderson moved backwards towards Vinny, standing ready in the suit, the handler moved forward encouraging his dog to "Get 'em, boy." The dog was crazy with adrenaline, pulling hard on his leash and dragging his handler along.

Anderson moved behind Vinny and gave the command, "Release."

The handler released his dog and both came charging hard, with the handler screaming at the top of his lungs. In the blink of an eye, the K-9 was behind Vinny and went for flesh just above the boot top. Vinny was helpless, at the mercy of the dog's handler and Sergeant Anderson to rescue him.

To his credit, Vinny fell to the ground, giving Sergeant Anderson a clear path to get the attention of the dog. The Sentry Dog was attached to Vinny and not turning loose. Sergeant Anderson was just as quick and attacked the dog with burlap sleeve and the switch. The dog released and went after the sleeve as the Sergeant was backing away.

The dog's handler moved in, secured the dog to the leash, and moved out. The dog was nuts by now and it took everything the handler had to move him away from a bleeding Vinny lying on the ground.

Vinny sat up, his green eyes still glazed from the shock of the moment. "I'll make no excuses why I fell, but the sun was in my eyes. I thought I had my balance, but I didn't see the dog until he hit the back of my leg."

Vinny was on his feet now and out of the suit, very thankful for Sergeant Anderson moving in and saving his ass.

As for the handler and his mighty dog, he couldn't be happier. We all wanted a badass dog that was workable. Any K-9 tasting blood was more alert and aggressive and made our future jobs in Vietnam easier.

Week eight is all about testing the skills we've been drilling into you," said Sergeant Anderson. "It's do or die time."

"There's a saying in K-9 that rings true, what goes down the leash comes back up the leash," said Sergeant Adamson. "If you're lazy, your dog will be lazy; if you don't pay attention to your dog's alert and act, eventually your dog will stop alerting."

Adamson's blue eyes searched the room for Airmen who weren't paying attention.

"You will be in a world of hurt on that night you have intruders and your dog didn't alert. There goes your edge."

"When your name is called," said Anderson, "bring your dog up on the choke chain to the starting position. Effect a changeover from choke chain to working collar. Next, find the direction of the wind."

He marched back and forth, changing direction each time he issued a new instruction.

"Put your dog on alert by giving the command, 'Watch 'em, boy,' and move out. Start patrolling and quartering your post, paying attention to signals your dog may give that he is alerting on something."

"OK, Moore, you're up."

My knees were trembling like some cartoon character and my palms were sweaty, it was my turn and I was ready. After the changeover, Duke and I moved out. "OK boy, it's show time, so let's move out. Watch 'em boy," I said.

From watching teams being tested I knew there was little or no wind, so I headed to the far end of the post opposite where the decoy was placed and began quartering. Still no wind.

"OK, Duke, we only get a couple tries at this, so let's get it right the first time." And off we went. "Good boy, Duke."

We moved through knee-high grass, with my eyes focused on Duke's head, nose and ears. I even watched the fur on his neck, I couldn't afford to miss the all-important alert.

Fuckin A, Duke, is it really fair to test me when I've never seen you alert? Help me out, buddy.

"Good boy, Duke, watch 'em," I whispered in my dog's ear.

We had quartered two-thirds of my post before Duke alerted, a strong alert, with ears turning, nose up, and head pointing towards the decoy. Then boom, Duke was moving and pulling me along.

Oh Lord, don't fall.

Duke was sprinting and it was all I could do to keep up. Sergeant Anderson disappeared behind the decoy who was standing in the suit, braced and ready to receive the entire force of an eighty-pound dog running at full speed.

I couldn't match his pace and Duke was about to pull my arm out of its socket. Boy, I was ready to let this fanatic go.

When I dropped the leash, Duke took off for the suited decoy at full speed, leaping and hitting the suit square in the chest. I got to Duke just as Anderson came from behind the suited man.

I yelled, "Duke, out, release!" Duke looked at me as I praised him up. "Duke, sit." He did.

"Stay." I moved out and walked the several steps to the decoy. As I searched the decoy, he knocked me down and took off running. Duke was on him before I got to my feet.

"Duke, out, release," I hollered.

Duke released and went to the sit position, real pretty-like. I was impressed.

"Praise him up! Moore," said Sergeant Anderson.

"Airman Moore, excellent job," said several of the Instructors who were hidden among the trees. Duke and I had performed and, on top of that, I was through. I had passed. Sentry Dog School was over for me.

On Wednesday, just before quitting time we were called together for our last briefing before Friday and graduation. All thirty-one of this class had passed. Tomorrow we would be moving our dogs back to Lackland and the kennels.

"My, how time flies when you are having so much fun, right men?" Sergeant Anderson said. "You've worked hard, you've had fun, and this training and our time together here will be with you for the rest of your lives."

A loud cheer went up from the gathered teams of men and dogs.

"We are letting you go early, that is, after you clean up the kennels, water and feed your dogs. Then stop by the canteen to grab a beer and some fried snake. So, get to it and have a good weekend."

"Airman Moore." Sergeant Anderson pulled me aside. "I need to talk to you for a minute before you leave today."

"Yes, Sergeant." He pulled me over to a group of Instructors who held bottles of beer and were very close to being shit-faced.

"Moore, we have selected you to be our demonstration team on Friday for graduation."

"No shit?" I was dumbfounded.

Sergeant Anderson and the rest of the instructors standing around broke out in laughter while I just stood staring at them, wondering what I was getting me and Duke into.

I laughed a little and shifted from foot to foot. "What will I be doing? Can you tell me, or is it an Air Force secret?"

"For your information, we picked you and Duke out of twenty-nine other dog teams. There was good reason for choosing you and Duke. Mostly it was Duke." Anderson smirked at me. My eyes were wide with questions.

"Moore, just be at the kennels at 0900 hours and we will run through the exercise. Duke has done this before, so by the time you finish, you and Duke will be the toast of Lackland. Don't worry, this duty will be fun, just don't fuck it up."

Plain Crazy

Joe was the only other team on the field. As I approached the gate, the Red Cross ladies with donuts were waiting on me and Joe. I turned towards Joe to hurry his butt up when he released his dog, no muzzle. Prince was at full gallop. Luckily, nobody was paying attention, including instructors or girls.

I froze! I put Duke between me and Prince for protection and held on to a donut.

"Prince, Halt!" Yelled Joe from fifty yards away as he calmly walks over to a sitting Prince, grabs the leash, "Hi Girls! Thanks for the donut." Winks as he passes, half the donut gone. We both saved a portion of donut for our animals. After eight weeks it was habit to bring a snack along.

I was on cloud nine. It was nice to be the center of attention on the bus ride back to the kennels.

Just think, Duke and I are going to demonstrate a real attack with teeth bared, no muzzle, full speed ahead, guns a-blazing, just like in a movie. It won't be Lassie or Rin Tin Tin attacking, it'll be me and Duke.

◊ ◊ ◊

Graduation, Lackland Air Force Base, San Antonio, Texas – early December, 1968

Thursday evening, after we had settled back into the K-9 barracks, Vinny and I were hanging around in the dayroom.

"I wonder what Joe is doing tonight?" I said. "It's been weeks since we've spent some party-time together."

"I'm with you on that," Vinny said. "I've asked him several times where he goes on the weekends and he wasn't forthcoming. In fact, he was vague to the point of getting fed up with me for pressing him." He tapped an emphatic drum beat on the back of the chair in front of him to make his point.

"Uh, well, what do you think he's doing? I'm sure it involves a girl or girls."

"I'd bet my last dollar on it, but he's close-mouthed about it."

I joined Joe for evening chow and it was the first time I had a chance to ask about his secret life.

"Joe, what's going on, my good man? Are you glad school is almost over?"

"Uh huh, and I'm sorry to see it end, though. I don't want to leave San Antonio. I like Lackland."

Joe lifted his glass and took a big drink of his Pepsi.

"But just think, in another two weeks I'll be home in cold-ass Jersey. A late Christmas with the family, snow on the ground, I'm at home, can't wait."

"I know what you mean but there will be no snow for me and I thank God for it."

"No snow and you call that Christmas?" Barry had just returned to the table with another helping of mashed potatoes. "I can't stand thinking about it."

"Joe, I'm dying to know what you do on your weekends. You seem to disappear every weekend since we've been in school. You've got a special girl in San Antonio?"

"Yes, I've b-been staying with her every weekend since Security Police School."

"Oh, my gosh, while I've been holding hands and stealing a kiss every now and then, you are shacked up with a beautiful lady." I blushed at the thought of my friend's worldly experience. "Does she have a car?"

"She has a car and an apartment downtown; it's a real neat setup. Man, I'll be sorry to see it end."

"Well, tell me about her, like, what's her name?"

"Her name is B-…b-…etty. I've been fucking myself silly since my first trip downtown after Basic when I met her at the Hemisphere."

"I think I should have hung around with you to find out how to get my love life in gear. Damn, Joe, good for you, man." I nodded my head in approval. "Vinny and me, we really missed hanging out with you, you scumbag. I guess I'm envious."

Friday morning came and Duke and I were waiting for Sergeant Anderson in the exercise area. "Do I need to put the muzzle on him?" I asked.

"Not now, but when you bring him out for the demonstration, he'll have to be on the leash, of course, with choke chain and muzzled. You'll be introduced as you come to about here, just inside the gate." Anderson pointed to the spot where Duke and I would be.

The Sergeant with the rifle and sleeve walked over to a makeshift bunker made from stacked sand bags that stood about four layers high. He then got down behind it and pointed the rifle towards where Duke and I were waiting. Sergeant Anderson explained the whole set up and there was nothing to it.

"The demo will come at the end of the speeches, and that will be it, graduation over. Enjoy your graduation ceremony, go ahead and sit in the stands and I'll come for you."

"Yes, Sergeant." I said. They turned to leave and I brought Duke back to his birdhouse.

The Base and Squadron Commanders arrived at our little graduation ceremony by golf cart with the Base Commander at the wheel. They were followed by security, also in golf carts. We marched in formation from our staging area to the grandstands. There was a crowd of happy faces waiting for us and we were proud of ourselves for the work we had completed.

"I would like to thank all of you parents, wives, and friends," said the Base Commander, "who have come to share in this graduation of Security Policemen who have taken a big step in their Air Force careers as Sentry Dog Handlers." He gestured toward the Airmen. "You should all be proud of yourselves. I know the U.S. Air Force is proud of you."

Much applause.

"Air Force personnel in Southeast Asia will sleep better at night knowing that you and your dogs are on duty. Now go forth to your next assignment and give it the best you have, first for your own sake, second for your families, and then for our country. God bless you and keep you. Thank you."

The Squadron Commander took the mic from the Base Commander as he left to get a good seat for the upcoming show.

"Ladies and Gentlemen, we welcome you to Lackland Air Force Base. We do not often enough get the opportunity to showcase our Sentry Dogs and their Handlers." The Squadron Commander beamed with pride. "We are honored to be able to demonstrate the capabilities of our highly skilled dog teams."

There was more applause, then I brought Duke and put him in the sit position in front of the grandstands while the announcer explained what the spectators would see. Vinny, Joe, Mike and Barry and the rest of my class were sitting together waving at me. All the eyes of some one hundred people, including the Base Commander and our Squadron Commander, were upon us.

Father, I prayed, *be with me now in this hour of need. Don't let me mess it up. Amen.*

I went through the gate and secured it. By the time I had Duke and me in position, Duke was already going nuts. The crowd loved it but Duke was doing a number on my arm and shoulders.

"Just get this muzzle off me and let me go," I imagined Duke's thoughts to be as I held him close and kept him in the sit position at my side. I removed the muzzle and let it fall to the ground. Good thing, because Duke tried to take off. What dog wouldn't want to go after the guy behind the bunker? But I held tight, got him back in the sit position and finished my changeover.

"Duke, we cannot fail, boy."

Now with the changeover completed, it was time to let Duke rock and roll. It was safe for Duke to do his thing, which was to growl a lot, run as hard as he could, and attack a bunker with guns blazing in the face.

I knew there was cheering and shouting going on behind me but it began to fade as I whispered into Duke's ear. "Watch 'em, boy, watch 'em."

Duke was at the end of the leash and pulling me so hard I almost fell. The Sergeant, who was acting as the bad guy, was jumping around waving his arms and the gun, shouting, just the things military working dogs loved to see and attack. Duke was so very ready and as he pulled me closer and closer to the bunker and the armed man, then I shouted, "Get him boy, go."

Duke was gone in a flash. The armed man began to shoot and continued shooting as my dog moved in closer. Duke was at full speed as he sailed over the sandbags, knocking the man flat on his back and sinking his teeth into the arm that was protected by the burlap sleeve. As the now unarmed man tried to keep Duke connected to the sleeve and not any of his body parts, I came up, grabbed him by the collar and shouted, "Duke, out. Release, boy. Good boy, Duke."

Duke didn't want to let go, but he was well trained and let go on the second command. We could hear the crowd roaring behind us with appreciation. He loved this. I was loving this. Even the Sergeant who took a fierce hit had a big grin on his face.

Duke got a "that a boy" and a quick scratch on the ear.

I got Duke to heel and sit and just held on to him so he could calm down. The crowd was quiet in anticipation. I was on autopilot now and I knew nothing was going to mess up my fifteen minutes of fame.

The Sergeant was standing with his hands in the air, still attached to the sleeve. I knew from experience that sleeve was getting heavy and I needed to get moving. I stood up, and using hand signals and voice commands I told Duke, "Stay. Watch 'em."

Thank God, the crowd is quiet because this is a critical point in this whole exercise. Will Duke hold and not attack? It's in his blood to attack.

When I moved forward, I caught Duke out of the corner of my eye and it looked like he was going to stay put as trained. He didn't move and I continued forward to my prisoner to perform a search. My prisoner turned and knocked me to the ground, and again Duke attacked on his own, locking his teeth into the thick burlap sleeve. This time he didn't want to let go.

It made for a good show but I had to get this maniac dog off my Sergeant before Duke decided he wanted to bite something else other than burlap. The crowd grew quiet again, sensing a very dangerous moment, and they were right.

I had no choice, so with the encouragement from the Sergeant to choke him out, I did just that. Two fingers to the throat and my squeeze cutting off the air flow should have brought Duke to his senses, but it wasn't enough.

Duke still refused to turn loose though I continued to stop the flow of air. He was showboating for the crowd and they encouraged him on. I held Duke's body and restrained him from going after anything but the sleeve, or at least I tried.

Once the need to breathe overcame Duke, he released his bite. I backed him off and he barked and growled at the Sergeant, then I put Duke in the heel position and gave him the command to stay and to watch him. Next, I completed my search of the prisoner, returned to Duke, took up the leash, and moved out with the prisoner in front of me. Show over.

The crowd gave Duke and me a big round of applause. We were walking tall.

After muzzling Duke, I marched him into the crowd. We were given plenty of room as the people backed away.

I was a little embarrassed by the attention I received but there was really nothing special about it, just having fun talking about dogs to people who cared about animals. I walked Duke to his birdhouse realizing this was it, my K-9 training was over.

I chained him to the birdhouse and started scratching him in all his favorite places. I filled his water bucket for the last time.

I am doing all this shit for the last time. But no, not true. I have another dog waiting for me in Vietnam. If he's half the dog that Duke is, I'll have it made. In five weeks, I'll be somewhere in Vietnam serving my country.

"Duke, boy, you are one of the best things to be happening in my young ass life. Thank you, my friend."

I grabbed him again and gave him a bear hug. He stood still as I released him, got up and walked away. He didn't bark or run around in circles about his birdhouse, he just stood and watched me walk away. He knew what was happening and he was sad about it, but dogs have a very short memory, at least that's what I'd heard. He would fall in love with a new handler, and love him just like he loved me. There was comfort in that.

Thank God, it was Friday night and the start of a long, fun-filled weekend was getting underway. Mary and I were having a fine time enjoying the music and good company.

Vinny had gotten off with a minor wound the day he fell, and it wouldn't slow him down one bit on the dance floor. He was floating on a cloud, his lean body attracting lots of attention from secret glances as he swayed to the music. He'd been showing off the gash on his ankle, telling them how he was a real dog handler now.

As the last chord to "When A Man Loves a Women" faded, the lights came on and we headed for the door.

Mary leaned in and touched my arm. "Will you accompany me to the bus stop, and ride with me to my barracks? I want to talk to you."

"Of course," I said. My mind was on Mary's lips. What came next wasn't unexpected being in the Air Force, but it still caught me off guard.

"Jimmie, I got orders today and I'll be leaving tomorrow. I have to report for duty Monday morning at Sheppard AFB in Wichita Falls."

"Is that still in Texas?" I asked. "Or maybe Kansas?" I sat stunned, gazing into her moistened green eyes. I had forgotten to breathe. "Well, you'll be missed, I can tell you that right now!"

I knew we had some good times together but she was moving on now, and maybe she didn't want any loose baggage slowing her down.

"Do you want to stay in touch?"

"I'll be missing you, too. Yes, let's stay in touch," said Mary

We fell silent as we made the loop to the women's barracks. I was again in deep thought. Two girls fought for my attention at the same time.

Lillian's sweet face with her light brown eyes entered my mind, and with it came the glare of lights, the sound of bells ringing, and an old station wagon that didn't look like a car anymore. All that was left was a jagged pile of twisted metal, razor sharp and bloody.

Why did I go with my cousin Glen to see the wreckage? Why did he ask? All I've got for my troubles are nightmares.

As hard as I tried in my imagination, I could not come to Lillian's aid. In my head, I heard the screams and felt the terror that must have been hers, and it would not leave me.

And here I am, not four months later, saying goodbye to another woman. I feel like a betrayer.

The following week in Asian Zone Readiness training (AZR), I enjoyed the two full days on the firing range, shooting everything from a .38 pistol to an M-16, firing hundreds of rounds for hours on end. I fell in love with the 12-gauge shotgun, I could hit anything with it. After my shoulder got used to the recoil I decided this was the weapon to use in Vietnam.

I thought the M-79 grenade launcher was going to be a lot of fun. Instead, it turned out to be a dud. We had to use dummy rounds that went pop. Same thing happened when throwing the hand grenade, but no pop this time, just a fizzling sound. Talk about unfulfilled expectations and plain ole embarrassment.

They did tell us when we're in Nam to always use rubbers. Something called the Black Syph was an incurable strain of syphilis that was so deadly, the military sent the infected men to an island where they could live out their last days in pain and dementia until they died. It scared the living hell out of us and we were not about to touch an Asian girl.

Imagine Mom and Dad learning that their son died in exile and was written off by the military. KIA by the Black Syph. Thanks, Uncle Sam.

The AZR instructors never once mentioned K-9s and the role they played as first line of defense. Maybe they spent too much time in a bunker or tower or riding around in a jeep.

By Thursday evening we gathered around to make ready to play war games, a version of capture the flag, in the dark. They gave us M-16s with plenty of blanks to shoot. Vinny and I found a small bunker in the trees, climbed inside and waited for the war to come to us. We stayed there for several hours until the games ended and never saw a soul. We fired off most of our ammo, returned to the staging area for a briefing, and then we were released to clean our weapons.

What a joke. GI is just another word for expendable.

◊ ◊ ◊

Starke, Florida, December 1968

US Headline News

US spacecraft Apollo 8 Enters Orbit around the Moon.

Astronauts Frank Borman, Jim Lovell and William A. Anders become the first humans to see the far side of the Moon and planet Earth as a whole. The crew reads from Genesis.

My Father picked me up at the airport for the hour ride to Starke. As usual, there wasn't a whole lot of conversation going on. After a few questions about dog school and how Mommy was doing, we both settled back for the rest of the drive.

I called Mary early Saturday morning, hoping I could catch her in the barracks. It took several attempts and several ladies to get her on the line.

"Hi, Jimmie," she said. Her voice was broken, throaty. "I've fallen in love with a Tech Sergeant here and I'm cutting off our relationship. It happened so fast and I know I'm letting you down. I'm sorry for that."

"Jimmie, are you there?"

Oh, I'm here alright.

Since I wasn't speaking yet, Mary continued, "Jimmie, thank you for the good times at Lackland. There is talk of marriage and I like this guy a lot. Will you ever be able to forgive me?"

"Whoa, Mary, that was a mouthful. There isn't anything for me to say, is there?"

"Please, Jimmie, please forgive me. Let me move on with my life and with peace."

"Mary, I could never be ugly to you…I do forgive you…and I want you to have a happy life…so yes, you are very much forgiven and you have my blessings for a long and happy life with your new man."

"Jimmie, I must go, I'm having a hard time with this and I'm sorry. Good luck. Keep your head down and your powder dry in Vietnam. I will be praying for you. Goodbye, Jimmie."

The line went dead and I sat holding onto the receiver, not wanting to put it down for fear of never hearing her voice again.

Mom walked into the kitchen just as I hung up. Her sad, blue eyes held compassion. "Sorry, Son, you don't deserve that, but damn… it happens."

"Thanks, Mom."

"Hey, try to enjoy the few days you have here with us. I'm making beef stew for supper tonight. Can you make it?" She fussed with a stray lock of auburn hair that had slipped from her barrette.

"You bet, Mommy. No way would I miss your beef stew," I said. "Mommy…thanks." We gave each other a big hug just as Daddy walked in.

"What's going on? Somebody die?" he asked.

"Well, sort of," Mom said. "Jimmie's new girlfriend dumped him."

After a few days of visiting relatives and friends it would be time to pay my respects at Lillian's grave site. In the back of my mind was the thought that I may never see this place again and I'd better get my personal things in order.

Heck, no wonder Mary was so uncomfortable. I am uncomfortable going to Vietnam. Maybe I should have gone to cook school after all… no, that's useless. Sentry Dog Handlers are elite and I am a part of that, I belong. Don't forget the edge.

Lillian wrote the most incredible letters anyone would want to read. All through high school she would have a one-or two-page letter for me each day before school started, folded up into a two-inch square. While I was home I would take the time to read the couple hundred or so letters Lillian had written me during our three years of dating.

The very first of her letters were a hoot, puppy love. It took most of a day to get through them. When I finished, I was dog tired and thirsty so I decided to get a six pack and visit her grave site.

Bradford County was a dry county, meaning no alcohol. There was a small country store at Kingsley Lake where we bought beer during our high school days and again I bought what I wanted, risking getting myself put in jail and the store owner put under the jail for the crime of thirst.

It pisses me off that I could be forced with the threat of jail time and at the point of a gun to bear arms for my country but can't drink or vote. I got the beer and made my escape, then drove out to the cemetery, thinking, *fuck the law.*

After the first couple beers, I found myself talking to Lillian's headstone, a recent addition. I was angry at myself for having joined the Air Force and leaving her behind. I never intended marrying until I returned from the war. I'd seen enough movies about war brides. I decided a long time ago to never leave a woman to wait on me while putting my life in harm's way fighting in some shithole country.

I should have knocked her up and gotten married. But that isn't my way and I have to deal with the facts; God decides when he will bring his children home. Lillian is with the Lord and that is that.

Still, it hurt like hell, and I was running out of beer. I had carried her last two letters with me since August but never opened them, and I began to examine them. The last one read:

> *Dear Jimmie,*
>
> *My Airman, I miss you something crazy, when are you coming home to me? This morning I told you how much I loved you, did you hear me? It's really, really cold this morning and I'm not looking forward to getting out in it. I've got to be at Carp's at 9:00 and I'm supposed to get some extra hours. Got signed up for classes at Santa Fe yesterday so a few extra hours will make up for that.*
>
> *I am so looking forward to seeing you, it's boring, boring around here without you to hold me and love me up. I miss you.*
>
> *Have you decided on going to Security Police School? That would be so cool. By the way I wish I had married you before letting you get away from me. I'm so lonely. I think I'm going on a date this weekend with Gloria and Janet. It will be nice to get out of the house and take in a movie. I can't stand you not being here. I'll have to dip my own popcorn in my own soda, that's not any fun.*

Gotta' go and get ready for work. If going to Security Police School is going to keep you from me one day longer, then you come on home. No babes, I understand it's a major decision. I'll have to be lonely for a little while longer. Miss ya; hurry home, your gal is waiting on you. I love my Airman! xoxo

All my Love,

Lillian

Lots of love and kisses when you get home, hurry.

I was in tears as I pulled away from the cemetery. I longed to hold her and would miss her letters over the next year. Her spirit, held close to my heart, was all that remained.

The Good Book says, "To be absent from the body is to be with the Lord."

I thought about that verse as I drove back into town. I was grateful that she would find rest in the arms of our Lord. It was the first time I'd ever considered what it meant to be buried in the ground. I rode around for a while before going home.

So damn final. Ashes to ashes are all that's left. It's time to move on.

On a freezing, crystal-clear Monday morning my Dad drove me up to Jacksonville to catch a plane. I was excited by my last-minute change of plans to welcome my sister Renee's brand-new baby boy, Craig Jr., in San Francisco before I took off for the war.

My Dad was driving, and as usual, was a little slow in responding to me, so I spoke anyway. "I was spending my last evening with Mr. and Mrs. Richards and just before getting up to leave Mr. Richards said to me, 'Now listen to me, meathead, a little advice from an old Navy man. You're going to be messing around with those native girls over there, so remember to use protection.'"

"That's good advice, Jimmie," Daddy said. "What upset you? Was it because of Lillian?"

"Well, yeah," I said. I was trying hard to keep my feelings under control. "I didn't expect that from him. I've dated his daughter for three years and just didn't expect that. But you know, come to think of it, he was telling me to move on, wasn't he?"

The only sign that Dad was listening was a small nod of his head, but I persisted.

"I hated hearing the words, but he was right. This part of my life has come to an end and it's time to get on with life." *So brave.*

We were pulling into the drop off lane for departing flights. "Jimmie," said Dad. "It's been a tough five months for you and going on to Vietnam despite it all, and as much as I hate that damn war, well, I'm really proud of you." The car came to a stop, engine still running.

"You know, Daddy, it seems all I've been doing since I got home is saying goodbye and tying up loose ends. Dog school was a powerful help in the moving on part, but I just can't fathom forgetting Lillian."

I waited for his response for several seconds and was about to give up and open the car door to get out when Dad spoke again.

"You stood tall through it, and Lillian's death must have been hard for you while facing Vietnam at the same time. I know how close you two were and I don't know if I could have handled it any better. Now see here, Jimmie, you go and visit your sister and have some fun with little Craig, go on to Vietnam, just remember to come home. Be safe."

I got out, retrieved my duffel bag from the trunk, waved goodbye and entered the terminal, leaving my Dad at the curb. He dismissed me with a wave of the hand, turned, got behind the wheel, and was gone. He never said, "I love you."

A week later I was aboard a plane to Vietnam. With no drinking on board the plane, the pressure was thick as glue, and no one was in a partying mood anyway. Here and there, a few played card games, but most tried to sleep. Some wrote letters for home-folk going in the other direction.

My God, I'm going off to war–isn't that special?

There were two hundred men of every rank and branch on board this plane and most kept to themselves. At thirty miles out and still over water, the pilot announced, "Any landings coming and going anywhere in Vietnam are considered dangerous."

All heads shifted upward towards the overhead speakers and then we eyeballed each other when it hit us, *We're here. Fuck-shit!*

"We know the stories of flying machines being shot out of the air using bow and arrow alone. There will be a sudden drop of altitude as we make our final approach to land and it could be a rough landing, so make sure you're buckled up."

From somewhere in the plane a voice yelled, "Now you're going to understand what pucker up means."

We fell from the sky like a rock. The engines cut back power to the point of an eerie quiet and the silence within the plane was like a vacuum. We held our breaths. I prayed. At one point, I thought I floated against my seat belt. Then bam! We hit the runway hard but rolled in a straight line with the engines in full reverse, and the big 707 with two hundred puckered asses arrived alive in Vietnam.

"This is your pilot speaking. Welcome to sunny Vietnam; we do that kind of landing every time I come to this place. Take offs are a little more difficult and without the pucker power, just climb, baby, climb. It's a nice warm day for your first visit and a comfortable 104 degrees. Good luck, men, be seeing you next year for the return trip. The crew and I thank you all for your service to our great nation and the opportunity to serve you."

I was a Florida boy and I knew about heat, but even I had my breath taken away as I went through the open door and grabbed for the railing to steady myself. Where was the air? By the bottom of the ramp my uniform was wet with sweat. We gathered to collect our baggage and then walked to the flight operations center for customs check of passports and to turn in any greenback dollars we might have. Seems the Vietnamese people prized U.S. currency for its backing in silver. In return, we got Military Script which was backed in blood.

Damn, it is hot.

My first night in Vietnam I would have slept naked, but concerns about getting to a bunker if we were hit during the night kept me in my underwear. After breakfast, we gathered our gear and caught the reliable blue bus to the terminal for the flight north to Phu Cat Air Base.

The C-130 was a brute of a workhorse for the Air Force. Fully loaded, this baby could land on just about any land surface, even beaches. Looking around the huge belly of this very loud and shaky plane I noticed something missing.

Excuse me, but there is a war going on. Where are the weapons?

The only weapon on board that I saw was a handgun carried by an Airman who was part of the crew.

If we go down in the jungle, he will protect all of us with his trusty thirty-eight and six rounds. Oh, hell, this is strange.

I looked down at my wet palms resting in my lap and realized I had taken up praying. I wanted a gun and bullets in the worst way.

Am I the only one seeing this? Ah, man, now I've got a headache.

I lucked out and had a small window where I could look out if I bent low and twisted around. I lost interest in the ground as it was green and stayed green, so I settled down to snooze a little bit. Soon, our plane taxied to a stop in front of a small terminal.

This must be the place – home to 5,000 souls who need protecting, my specialty.

Chapter Four

Welcome to the Cat

Phu Cat Air Base, Vietnam – January 1969

US Headline News

Ohio State Defeats USC in Rose Bowl to Win 1968 National Title.

The Beatles Give Last Public Performance on Roof of Apple Records.

Nothing in Vietnam was built to last, everything seemed to be built of wood. The terminal building had air-conditioning and as we passed through the building I noticed in a corner a small counter where you could order hamburgers, hotdogs and ICE CREAM, holy moly! Ice cream. I would check out this place soon enough.

All nine of us mounted a deuce-and-a-half truck for the ride to our new home away from home. None of us had any idea what to expect, but we saw paved roads, large buildings of all shapes and sizes, two-story barracks. Heck, it looked a lot like Lackland from that distance. Maybe that was the point.

We pulled to a stop in a sort of a cul-de-sac with rows of barracks on three sides. The barracks were two-story, stick built structures. There was green grass that tried to cover the rocky red clay that lay everywhere. Two guys on mowers were cutting the grass.

I pointed at them. “Hey Joe, do you think they get combat pay?”

He laughed and stuttered. “I don’t know about them but if I get killed by a flying rock I hope it won’t affect my insurance.” Our group laughed as we dismounted.

The side construction of the buildings/barracks were of wood slats set at forty-degree angles for ventilation with screen netting to keep the bugs and rain out. Nine of us followed an Airman in the direction of the barracks.

“Hey, you guys K-9?” Someone shouted at us from the second floor of a barracks off to our left.

“Yes,” Barry said. Now there were a dozen or more guys hanging off the railing, watching and yelling like we were on parade.

“How many?”

“Nine,” we shouted back. A cheer went up as more and more of the guys were filing out of upper floor rooms to look at us. Some were near naked, running around the porch shouting obscenities.

“What the hell is going on?” Vinny cocked his head sharply over his shoulders. “And why are those men hollering at us?” Vinny asked the Airman who was taking us to our quarters.

“K-9,” he said.

The Airman led us into a barracks and assigned us our beds and lockers. He gave us directions to the base laundry where we would get sheets, blankets and pillows.

“This is just temporary until they make room for you to move to the K-9 barracks. Don’t ask how long it will be before you move, could be several weeks or a month, I have no idea.”

He pointed out the door. “Airmen’s Mess is that way, can’t miss it. You are to report to the pickup area out in front of the mail room or where you got off the truck at 1300.”

Not wasting any time, are they? Probably give us a couple days to get our dogs on leash and hopefully we’ll have weapons issued before too long. Imagine, sheets, pillows, and blankets. Blankets? For what?

“Save your questions for this afternoon when K-9 people will pick you up for the ride out to the kennels. They will get you settled in, which will take the rest of the day. Any questions? Good luck in the Nam." He turned and walked out the door.

The Bear flopped down on a mattress, exhausted. "They sure seem glad to see us," said Barry Teagan.

We stood around for a minute or so trying to make up our minds what to do next.

"Where you guys from back in the world?" The call came from someone hanging on the railing.

From another, "Damn glad to see all of you."

When three Vietnamese women walked through the barracks' door, everything got quiet. Then a naked man walked out of the shower past the women, while we looked at each other in shock with gaping mouths.

"Can you believe we will be sleeping in barracks in beds with sheets and a shower down the hall?" Vinny mumbled still staring as the women disappeared into the shower room.

"Thank God I didn't decide to follow in my Dad's footsteps and become a Marine," said Russell Puls.

Even though the barracks appeared like they would fall in a bad storm, on the inside they were built very well. Concrete floors, wood frame, no windows, no air-conditioning. The exterior walls were covered in one-inch by eight-inch wooden slats, slanted at an angle up to the roof, two-stories high allowing for air to get through. Screens covered the inside area from top to bottom. If the door stayed closed, the theory went, fewer bugs would bite us as we slept.

We finished making our beds, then headed through the back of the barracks for the mess hall. We passed the three mama-sans standing over two guys skinning a big ass ole boa type thing in the shower. "Good gravy," I slowly whispered. "Oh gosh almighty knows, the size of that thing! Is it dead?"

Our path took us past several more barracks and then onto a most absurd sight. Construction on an Olympic-size pool, or close to it, was almost completed. We stood and marveled at our luck. Our barracks was only several hundred feet from the biggest pool in Vietnam. I felt cooler already.

"You got to be shitting me, in Vietnam?" said Barry Teagan. "I can't swim."

"M-maybe you can learn and get a job lifeguarding," said Joe. He dropped his cigarette.

It was then I noticed a large upside-down V of a building. "What the hell is that? Maybe it's the base theater?"

Food would wait as we all wandered over to check out the building.

"It's a damn church," said Vinny.

"Watch your mouth," said Joe.

The sign said this Chapel was a duplicate of the Chapel at the Air Force Academy, smaller in size, but everything else the same. Inside, it was simple in its wood framing and that made it pleasing to the eye.

We continued to the Airmen's Mess Hall, which was laid out kind of like a peace sign with no circle around its edges. It had the same concrete floor and wood frame construction as the other buildings. The building was also used as the Airmen's Club after the evening meal.

"No expense spared for the men who do all the hard and dangerous work in the force," said Barry. We all agreed as we got in line for our first meal at Phu Cat Air Base, Vietnam.

I took the good food as a blessing, my stomach being very important to me.

"When will we ever get the chance to see any night life and hang out with the day pukes?" said Joe.

After the meal, a few of the men began to wander off and someone yelled out to be at the pickup point at 1300. The rest of us gathered for a bullshit session that included Vinny, Joe, Barry, Russell our California surfing nut, and me.

There were Vietnamese women running all over the place in their black silk pajamas, sandals and their Ho Chi Minh hats, talking ninety miles an hour in their sing-song language and making no eye contact, much less talk. But no "round eyes." I'd never heard of calling women from the West "round eyes" before.

I wonder who the enemy is. Why wasn't I issued a weapon back in Seattle or as soon as I departed the first aircraft that brought me here?

The nine of us loaded up for a ride on the back of a deuce-and-a-half. On the ride to the kennels we passed fuel storage and bomb storage facilities and I couldn't help wondering why everything looked so wide open and vulnerable to attack. In fact, the base looked like someone had left the gate open. We passed a few bunkers and towers, but all were empty except for one tower. We looked off base, saw traffic moving along a highway, and a few small villages dotted the area.

At least the fuel and bomb dumps had chain link fencing around them. The perimeter fence line, what little there was, needed repairing, replacing, updating or something.

"Joe, quick, look, just a broken-down fence line."

Joe tossed his cigarette in the direction of the fence as it meandered off to who knows where. "No man, no fucking way. That's not a perimeter fence, can't be."

We could hear the dogs barking, all sixty plus of them. The truck pulled into the kennels and backed up to the ramp.

The sentry dog section was located at the top of a gentle, sloping hill that seemed to go on forever, ending in dark green mountains way off in the distance. The air base was set in a huge valley surrounded by spectacular mountains of every shade of green possible.

In the valley, amidst former rice paddies, was our big-ass air base. An empty guard tower located just outside the chain link fence that surrounded the obstacle course and the kennel area broke up the view around the kennels. There were sixty-five dogs housed in the kennel building with seventy kennels.

Several NCO's (non-commissioned officers) walked out of the building and had us fall in, then called us to attention.

"At ease men, my name is Master Sergeant Johnson, the Kennel Master in charge, and boys, we are so glad to see you Airmen show up."

The Kennel Master stood over 6'5" tall, easy. He had jet black hair, black eyes and a big smile. "We've been shorthanded for months and the wear and tear on the handlers and dogs is taking a toll. We will be moving you through orientation as quickly as possible to get you out on post and working. Our men need some time off."

No beer belly on this man. *I think maybe he is a long-distance runner, has that slim, almost too-thin look about him.*

"You will be issued your weapons and taken to the firing range first thing. By the time you get back to the kennels we will have your gear ready and issue your dogs."

Master Sergeant Johnson let out the loudest, "shut up" I ever heard. And I'll be damned if the kennels didn't grow quiet, then quieter, until the First Shirt was able to continue.

"Men," said Johnson. "Keep the muzzles on until you feel comfortable enough to work them without muzzles. I'm sorry you will have only two days to get to know your dogs, but you'll have plenty of time on post."

On the way back from the range we stopped by base supply to pick up our uniforms and gear: two sets of green utilities and two sets of camos, two sets of boots, web belt, two canteens, poncho, helmet, bug spray, and K-Bar (knife).

At the kennels our dog gear was laid out for us. We cleaned and put away our weapons. We spent the next half hour oiling up our leather muzzles, collars, and leashes.

It was time to meet the K-9s who would be our partners for the next three hundred and sixty-plus days. Vinny got King, Joe got Jackson, and I couldn't believe it – I got another dog named Duke. I decided to call him Duke II because he even looked like Duke the First, having the same build and color.

"Airmen, listen up." The NCOIC Master Sergeant Johnson walked over to us. He was the Kennel Master. "You go on duty tomorrow night with an experienced handler who will show you the ropes. You will have tonight off, but do not leave the kennels until you have taken your dog for a run and started the process of getting to know him. Tomorrow morning you will report at 0900 hours and continue working with your dogs. Any questions? OK, you're dismissed."

Duke II was barking just like the rest of the dogs, but when I began to clean his kennel, he stopped. His ears drooped down and he began to wag his tail. I started talking to him and he began acting like he had known me all my life. I held out the muzzle and leash to him and he seemed to relax, so I took the next step or two forward without any aggression from him and put the muzzle on him first, I followed it by the choke chain and leash.

Hot damn! I can't believe how easy it is to get the muzzle on Duke II.

Duke II was tall but stout. I bet he weighed close to 80, maybe as high as 90, pounds. He was playful, had perfect ears, light brown eyes, and had a missing canine tooth. Duke II's chest was white and yellow. Black and red/orange covered the rest of his body.

We entered the deserted exercise area, started at a slow trot and built up to an all-out run. I removed the muzzle and started to walk the obstacle course. He climbed, he crawled and he jumped. He right turned, left turned and about faced as well as the first Duke back at Lackland.

Voice command or by hands only, or on his own initiative, my dog hasn't missed a beat, but something does not sit right with me. Don't be silly, Jimmie. "Can't compare him to Lackland Duke."

The other handlers began to enter the secured exercise area and Duke II didn't react. Within the hour all nine of us new guys had met and overcome our beasts, the fear of a bad news dog passed as it seemed we all had lucked out. Duke II wasn't the only dog in the yard that didn't pay any attention to the other dog teams. All the dogs were happy to be out of their cages and we ran and worked them hard. To a man, we had accomplished the first order of business the First Shirt had given us. We called it a day and headed back for some supper.

On my first night at Phu Cat AB there was a gentle wind blowing across my body cooling my naked skin as the heat drained from the day, but the sweat remained.

If I get a chance to go to the BX, I'll get myself a fan, but for right now I'm taking it easy. No telling how long it will be before I get another night off. Some of the handlers have worked seventy and eighty nights straight without a break. I could see now why we were so welcomed on our arrival yesterday.

Early in the night, around 2300 hours, there came a loud boom of a big gun going off. We ran out the front door to see if we were being overrun and where the nearest bunker was. Someone back in the barracks yelled to us that it was an outgoing round, probably flares from big guns next door at the Korean Camp.

After spending the next morning at the kennels as instructed, I had some time to kill so I checked out the Base Exchange. I never saw so many electronics, watches and cameras in one place at one time. They even sold several sizes of fans and small window air-conditioners, radios and TVs. Everything was made in Japan. You could buy a car; a suit of clothes made in Hong Kong and plan our R&R right here. I wanted a new watch and camera for sure, and I would be back. I left with a pair of drumsticks.

FNGs

There were no introductions at Guard Mount that night. Thirty-one handlers waited in three lines for the Non-Commissioned Officer in Charge (NCOIC) to come out and give us, the fucking new guys (FNGs), our first nightly intelligence briefing.

The night boss, Staff Sergeant Francis, was fat and flushed pink with alcohol.

"Attention," yelled Francis. "At ease, we finally have nine new replacements and will start letting you guys have a night off just as soon we can, keeping in mind the rotation."

A cheer went up from the other handlers while we newbies stood around with grins on our faces.

"Okay, at ease, men. I've a lot to cover so let's get on with it. Our new people will be going out with some of the older guys tonight but without their dogs. Tomorrow night they will be working a post with a dog and at least nine of you will be off tomorrow night."

Again, a cheer went up from the handlers.

"You trainees listen to these older handlers. You never know when something they teach might save your lives. You guys teaching these FNGs, we need them up and going and ready to take over a kilo post tomorrow night."

"Situation report. The following Air Force Bases were hit last night: Nha Trang, Tan Son Nhut, Da Nang and Cam Ranh Bay; no WIA, KIA or MIA. Enemy losses at five KIA and one WIA. One C-130 received minor damage. Sapper teams and mortars were the weapons of choice. Second Corps was mostly quiet last night. Local intelligence is saying the same thing with some movement of a NVA regiment to our north. The Air Force has gone twenty-seven days without a casualty. That's good news, but don't relax, men. The enemy is always ready and waiting to cut your throats. Any questions? All right, let's load up and get to work."

My first night on a kilo post found me on the notorious north side. I had the good fortune to be teamed up with Sergeant Lott, who soon would go home.

Lott was a beefy guy, and his muscular dog Thor was built a little like him. He had a little scar under his left eye where he said a dog nicked him when he was wearing the burlap suit back in his training. As our truck crept down the orange clay road, Jack Lott pointed out the bomb storage area.

"Jimmie, when we get hit, we get hit right here, every time. Somewhere between Kilo Three and Kilo Ten, nowhere else, not even the main gate has been attacked. So, this ground is very important to the gooks."

"If the gook sappers blow either or both of these two sites, they could take us out of the war," said Lott. "Not for long, but long enough to stop the air cover we provide. Our fighters support a lot of Marines and Grunts on the ground, north of here all the way to the DMZ."

The red clay road was as smooth as asphalt and followed alongside the main runway, straight as an arrow. But the truck crawled along in second gear until we arrived at the assigned post.

Out here if we wore the right clothing it was possible to disappear and be swallowed up by the night.

"Remember, everything looks different in the dark."

We both dismounted the big deuce-and-a-half and I watched as Sergeant Lott first locked and loaded his CAR-15. Then he changed over his dog from choke chain to his working collar and removed the muzzle.

"Airman, are you locked and loaded with that damn weapon on safe?" he said.

"Yes, Sergeant, I'm on safe and ready to go."

"You stay about six feet behind me and be as quiet as you can so as not to distract my dog."

"Yes, Sergeant," I said.

I don't trust that dog. I'm not about to distract him.

"Airman, what's your name?"

"Jimmie Moore."

"And don't call me Sergeant, I'm no damn lifer. My name is Jack."

"OK, Jack."

"Watch 'em boy, watch 'em." He whispered in his dog's ear.

We walked off the road into darkness and I immediately fell on my ass, tripping on something before walking more than a few steps. Jack snickered and continued. There was no moon that night, or not yet anyway, and we were soon swallowed up by jungle. Jack and his dog moved about, but he laughed at how often I stumbled and fell.

"Airman, by now the whole damn country knows I'm out here with a FNG. Try and stay on your feet, man."

"I can't see shit; how do you walk in the dark?"

"It will take a while to learn. One secret is keep an eye on your dog's ass. Let your dog do the work for you. Take baby steps, try to be light on your feet, and that will cut down the number of times you end up on your ass."

We walked for what seemed hours and I fell many more times trying to keep up. Unstable and feeling miserable, I had already hurt one knee and twisted an ankle. I could not see Jack's dog though I stayed as close to him as I dared. Every twenty or so steps Jack would have us stop and he would just listen. His dog listened too. They worked as a team.

Lott stopped and told Thor to sit, then turned to face me, his nose touching mine. "Listen to me, Airman. Don't ever, ever assume the VC have taken the night off. The VC can lay as still as a cobra waiting for its next prey, waiting for some dumbass handler to screw up, like falling down every other step and giving away our position."

At least the Sergeant was smiling or seemed to be, and he wasn't finished yet.

"If your dog sniffs out the VC, the enemy will open fire with hell-fury. Learn your dog's alerts. Is his alert on human or animal? Snakes are a totally different thing. When your dog takes off running, just follow and get out of there, run."

With a quick turn of his head, the conversation was over and a hand was on my chest. Lott dropped to one knee, pulling me with him. "Quiet," he said.

We listened.

"The enemy knows that getting past the dog is going to be tough. The consolation prize, take out a team, especially the dog. Our dogs haven't allowed an attack to succeed, but they keep on trying. Be a patient listener."

I was drip-soaking wet and hotter than I had ever been. Lott watered Thor and stood motionless as he listened to the night. Few stars were shining through the cloud cover, making the night shadowy and still.

"Jack, I have to work a dog tomorrow night, you know, move around. How do I do my job when I can't see a thing?"

Lott was sitting on his steel five-gallon bucket and Thor lay at his side, furthermost from me still drinking water from a canteen cup. "Airman, I've got dog faith, the dog kind of faith. And you will too. You're crazy if you don't."

Lott rose, stretched to his full six feet, and grunted, nose to nose with me again. "The first day of school what did the instructor tell you? 'Trust the dog.' Jimmie, I'm telling you the best advice you'll get in Vietnam is 'Trust your dog.'"

I heard voices in the night but was not sure which direction they were coming from. I came to a sudden halt before handlers gathered in a circle, talking, smoking, waiting for the coffee truck to show up. My eyes had adjusted enough to make out two other handlers, their dogs and two new guys. Six of us were gathered in this small clearing with enough distance between us to keep the dogs apart.

I was tired and sweaty and ready for a break. Jack watered his dog and I walked over to the other new guys to find out who was who. *Joe and Barry, oh good, I can't think of a better bunch to hang out with.*

We FNGs sat on the ground while one of the handlers sat on his bucket, which meant we were on his post. The other two handlers left buckets on their own posts.

"You guys need to get buckets. They're a life saver out here," said one.

Jack spoke up next. "We carry our flak jacket, helmet, extra ammo and c-rations in these buckets. Your bucket keeps you off the ground and away from the ticks, bugs and snakes and all sorts of weird, nasty biting shit."

"These buckets keep your ass dry and out of the filth when it's raining buckets," one of them said.

We all had to laugh at the pun. I would have my own bucket this time tomorrow night. I didn't know what to ask, so I sat rubbing my swollen ankles and fighting off mosquitos.

"Watch out for Sergeant Francis, he likes to come on your post, try and catch you sleeping on duty. By the way, sleeping on duty is a firing squad offense over here. Don't get caught."

I couldn't believe my ears. "You mean there are handlers who go to sleep on post, in Vietnam. That's crazy!"

"You'd think someone would shoot the bastard," Barry said. "Has he ever caught anybody fucking off?"

"Not that I know of," said one of the handlers. "Just be careful and never turn your back on Charlie or a lifer."

Jack spoke as he got to his feet. "Out here you are on your own. Ask the driver who drops you off what he knows about a specific post and go from there. Tet is coming up in less than a month and there will probably be a truce. Last year's Tet was a real calamity. Every base in Vietnam was hit, so be ready and keep asking questions. Make sure you keep your dogs watered and watch for snakes."

Another veteran handler cocked an ear and announced the approach of the coffee truck, then continued. "Your dog will tell when there is a snake or a gook in the area. There is a difference in alerts, just pay attention."

It was a little after 0100 hours when our group meeting broke up and I was getting sleepy-eyed. I struggled to keep pace with Jack and his dog. The weapon I'd been hauling around was getting heavy. My aches and pains didn't amount to much. Though I wanted to sleep, it sure wasn't worth going before a firing squad.

The first thing I did back at the kennels and off duty was to find myself a bucket. Dog food was shipped in buckets and handlers extended their life by putting them to good use. All the new guys were busy getting their buckets squared away with more ammo and magazines. We could carry all the ammo we wanted, so the bucket carried our extra rounds. On our web gear, we had room for two canteens and four magazines. The rest went in the buckets.

That morning, when I returned from guard duty, a card game had already started and someone had acquired cold beer. It was 0830. After a few hands of five card stud and a couple of cold ones I was ready for sleep and I slept hard, but in a pool of sweat. You couldn't do anything without breaking out in a sweat. Boots would turn white from salt being sweated out.

Last Off, First On

I was working the late shift tonight and I would be manning Kilo 53, which was located at the east end of the base, with my K-9, Duke II. We were sitting at the front of the truck because I would be getting off last.

Last off at night, first on in the morning. The deuce-and-a-half carried ten handlers and their dogs, crawling down the road to its last stop, Kilo 53. I passed a tower, then a bunker and then we went through a fence line.

Oh, shit. Where are we going?

The further we traveled, the harder it was to see anything in the thick fog. I heard a radio playing from inside the cab. I looked over the driver's shoulder at what he saw and the fog looked like milk, it was so dense. The driver could not see past the dimmed lights at the front of the truck. The hair stood up on the back of my neck as I began to realize how dangerous my situation was turning. The driver knew where he was going and started to circle around, then he came to a stop. I dismounted and walked around to talk to the driver.

"How the hell am I to walk this post if I can't see a damn thing?"

"Look buddy," the driver said, "you'll get used to it in time, remember to take your baby steps. You'll learn real fast."

"What's your name, Airman?" he asked.

"Moore, Jimmie Moore," I said.

"This is your first night out with your dog and they assigned you to this post? Moore, you are at the ass end of the base and close to a mile outside the perimeter or the nearest fence line. The towers and bunkers are even further back and well behind the fences. If Charlie doesn't kill you, one of these damn fools in the bunkers will." He spat, irritated.

"Where are you from back in the world?" he asked.

"Florida," I said.

"I visited Daytona Beach with my parents. It's been a few years. I loved the beach, but hated the heat. It was so damn hot. Look Moore, you got a tough post to navigate tonight. Try to locate the other handler when the coffee truck comes by after midnight. Don't go looking for him. That can get you killed out here. Wait for the truck."

Just before he started to pull away he asked which of the dogs I had been issued.

"Well, I'm calling him Duke the Second because the dog I trained with back at Lackland was named Duke."

"Duke." It was almost a whisper as he repeated the name. "Oh shit, Moore, you be careful and go slow tonight, this is no damn race."

The truck circled me. Its taillights became invisible in a moment. And just like that I was alone. I could hear the truck make its way along the beaten road that led to the runway berm road and to safety.

I locked and loaded my CAR-15 and remained seated on my bucket and removed the muzzle and changed over Duke II from his choke chain to his working collar. I could not see my hand in front of my face, much less my dog, only by touch did I finish the changeover.

Fearing to move even one inch and not moving at all in the last hour or so, I remained seated on my bucket with Duke II at my side. Both of us were panting in the damp heat trying to breathe. As my heart settled down, my mind searched for anything that was comparable to this God-awful fog that choked me.

I remember when I was a kid running behind the mosquito-spraying truck with its thick, white poison. I was having fun playing hide and seek with my sister and disappearing into the sticky, cool mist the county truck sprayed. I laughed and remembered where I was, Central Highland, Vietnam.

I had made my radio check over an hour ago and still I sat, listening.

"Come here, boy." I whispered to Duke II, pulling him to me. I sat stroking Duke's already wet coat, not sure how to proceed.

I can't see shit, but I have a post to clear and it doesn't matter if there is fog, rain, or an enemy lying in the grass waiting to dispatch me and my dog. Not only do I not know how Duke II alerts but seeing any alert in this mess is impossible. Drowning in this coffin of dewy wet fog, I couldn't see Duke II's ass if my life depended on it, and it does.

I got up and told Duke II to "watch 'em boy." Then I took my first step into total darkness. I extended my weapon hand, and holding the small CAR-15 out in front, moved forward trying to touch a bush or tree or anything that would give me some sense of where I was.

Looking at the ground for any signs that would help me out, I found tire tracks that told me I was still in the drop off/pick up round-about, so I kept moving. The fog never let up, but I found the edge of the jungle and moved to the left using my hands and body, feeling my way as I went. Duke II wasn't much help to me nor did he ever take the lead.

Please, Mr. Custard, I don't wanna go, I imagined my dog saying. It felt like I was pulling him along while trying to find a hole or a path through the grasses and brush or whatever it was I couldn't see in this white blackout.

I had been on duty almost three hours when I came upon a path that changed my direction away from the round-about and into what I guessed was jungle. Stumbling and falling, making more noise than

Indians on the war path. I moved forward and deeper into what, I didn't have a clue.

After what seemed hours, I knew Duke II needed a drink of water, so we stopped to rest. Even though I could not see him very well I could hear him panting. I reached down to his mouth and with a touch I could feel the heat of the dog's extended and very dry tongue. He needed to cool down.

While Duke drank his fill of water from my canteen cup, I opened a can of peaches with an opener called a P-38. It was my first introduction to c-rations.

A P-38 Lightning was an airplane used during WWII. Comical how the military can reduce a beautiful aircraft to a simple little gadget like a can opener. Oh, these warm peaches are a treat sent by God.

We were moving at a snail's pace, but I was satisfied with that. The fog had not let up one iota and it was getting close to coffee time. There was complete silence in this wet stillness. I tried to get Duke II to move in front and take the point but he was having nothing of that.

I look forward to the return of the truck and someone to talk to. Dammit! What's going on?

The sky lit up and I hit the ground. My weapon was on full automatic and ready to kill something. Duke II sat and watched me. My night vision was useless and I froze as the bright lights pounded me. When it happened, I had lost my sight in a flash, damn blinding strobe lights. I found myself clinging to Mary on the dance floor.

My mind righted itself as I lay frozen on the wet, slippery ground and remembered where I was.

This is a combat zone, get your shit together, Airman.

It only took a second to realize that I was underneath landing lights that guided aircraft to the runway. Still disorientated, I got up on one knee and I could hear the beginning sounds of a whirling jet engine. Out of the night came the almost soundless rush of a ghostly F-100. It was thrilling the way its pointed nose cut through the fog and was gone. I stood up as the heat of a dying engine passed fumes over me.

The pilot turned on his landing lights at the very last second before hitting the runway, so the VC wouldn't shoot him and his plane out of the sky.

I wonder if by chance he might have caught sight of me saying my prayers and will have a good laugh at the Officers Club.

All the while Duke II sat looking at me like I was crazy.

I feel like a damn fool, but no one will ever hear about it. A lesson learned.

What a rush I felt as the second F-100 flew over my head and again, standing now, I was in awe. A soft, red warmth lit the cockpit and I could have sworn I could make out the pilot's helmet in the glow. I felt the hankering to duck.

I sat on my bucket and went to my c-rations for some crackers and peanut butter. Duke II turned his nose up at the peanut butter.

Are you kidding me? I never knew a dog not to like peanut butter. That's okay, buddy, more for me.

I decided to remain seated until the coffee truck came around. I was not a fan of coffee but I hoped to run into the other handler on Kilo 54. I was desperate for company and for someone to tell me about this fogged-in, mysterious post.

It had to be close to 0200 hours when I heard the deuce-and-a-half crawling its way to the round-about. I had not moved off my bucket for hours. For the first time, tonight Duke II took the lead and all I had to do was follow his ass. The truck had cut its engine by the time I made it to the round-about. Thank God, the other handler was there.

We greeted each other as he moved out of the way and let me approach the large, shiny coffee urn for a cup of Joe. That's what my Daddy called coffee in the Navy. I never drank coffee and didn't know if I wanted cream or sugar or just black. I wanted some help staying awake and straight up black would probably do the trick. With my cup in hand I moved to the other side of the truck, where I could be a part of the conversation now going on with the other handler and driver.

I could hear them talking about me. "Suicide post, and he shouldn't have been posted to it. Not everyone is like you, Rev," said the driver. He dismounted and moved into the brush for relief.

I came around the truck to where the other handler was standing and asked what they meant by "suicide post."

The driver had returned from the bushes and was making ready to pull out. "I told you it was the ass end of the world and it's close to a mile to get back to safety," he said. "If Sir Charlie decides to hit us on this end of the base, you better put on your running shoes."

"Hey, there," said the handler drinking coffee with his dog just a short distance away. "I'm Sergeant Dennis Collins, but everyone calls me Rev. The fog makes this a pretty spooky place to be at night."

In a way, Rev reminded me of Howdy Doody with his big ears and round face. His blue eyes and the pockmarks on his face gave him a look of maturity.

He shook my hand. "I love these two posts. I like being able to disappear. I'll show you the sights. Don't listen to any suicide crap about Kilo 53 and 54, they are my two favorites."

"Dennis, Rev, my name is Jimmie Moore and am I ever glad to see you. Is there a chance of you telling me a little bit about these posts? I've been unable to get off my bucket to even clear my post. I've been on my butt, tripping and falling more than I've walked."

"Moore, you're doing okay and when we get through with our coffee I'll take you on a tour. It really has a lot going for it."

"Thanks," I said. I threw the foul-tasting coffee to the ground. "How does anybody drink this stuff?"

"Get used to it. The mess hall doesn't bother to make a fresh pot of that shit for K-9," the driver said.

"Well, that was my first and last cup. I was hoping to use it to keep me awake."

"Let's head out, the fog's clearing and I'll be able to show you around."

It was a welcome relief to be able to see stars for the first time that night. I don't think I had ever seen the Milky Way this big before. A crescent moon was coming up.

Rev was a friendly guy and we enjoyed the rest of the night in each other's company. His dog was as big as he was and probably weighed more than Rev. His dog Rip sat contented next to Rev with one eye on me.

When Rev told me he was doing his third tour, I nearly fell off my bucket. He said he would finish his four years by staying in Vietnam and away from stateside foolishness. Anybody else would have said "bullshit," but the Rev said "foolishness." Rev would correct my foul language several times that night.

"My daddy's a preacher back in South Carolina. I want to follow in my father's footsteps," the Rev said. "I don't smoke or drink, and I love the Lord with all my heart."

The bright stars made following Rev and his dog around a lot easier. Kilos 54 and 53 were heavy in grasses and bamboo, the same as on the northeast side of the base but denser.

Rev pointed to his left and right. "There's no jungle anywhere on Phu Cat. Everything was bulldozed when they built the base."

Rev is right. There is a lot to like about both posts. From here on, I will look forward to being assigned to either.

I learned from Rev that night about the presence of ROK Army of the Tiger Division who protected a well, used for potable water located just out in front of Kilo 54 and 53. Their main camp called Thunderbolt butted up against the air base.

"The Korean Army stationed around the base are in place to protect us from attack. They've been there since the first days of construction," said Rev. "These 'Tigers', as their division is called, will set up ambushes and be out in the bush for weeks at a time."

"They sound pretty bad-a--." Out of respect, I stopped myself before I uttered the curse word.

"We didn't get to see any of this but we heard the rumors. You don't mess with these people. They also have a third, but smaller, camp located less than a klick north of the base."

"How far is a 'klick'?"

"A klick means 1,000 meters, one kilometer, or .62 miles. So, if an Airman radios, 'We're 10 klicks south of your position,' that means they're 10 kilometers or 6.2 miles away."

As morning approached, Rev said goodbye and left to meet the pickup truck when his shift ended. I was left alone to walk anywhere I chose to go and felt in control. I was nineteen years old with a weapon in one hand and a weapon in the other hand. I was in control, sort-of.

This is my little patch of ground and I am the boss. I've never been up all night before. Very strange. Keep moving around to get the blood going. Who would ever think about sleeping in this place?

Duke II didn't seem to focus for long when alerted, and wouldn't walk the point. In fact, he either hugged my leg or walked behind me. There was little sniffing the air or head or ear movement.

I was first on the relief truck at the end of the late shift.

My first night of duty as a Sentry Dog Handler was in the books, 363 to go.

By our third week we moved from temporary barracks to what was known as the K-9 barracks. All sixty-two handlers were crammed into the second floor. We had one latrine with four heads, four open showers, and a washer and dryer.

There were sixteen of us to a room. Joe, Vinny, and I managed to stay together. In fact, eight of the nine of us from Lackland were in this room. It was a small room for sixteen men to live in, divided in half by lockers placed back to back, eight men in one-half and eight in the other.

On our side, there was a large table in the center and this became our focal point for eating, drinking, card playing, letter writing and plain old conversation. Oh yes, and it was the TV and music room. As cramped as it was, it was home and these sixteen men became close friends to the point of seeking each other out while on post.

Heat Kills

At guard mount Sergeant Francis informed us, "Okay, men, listen up. Overnight we lost a Sentry Dog to heatstroke. Yes, it is a sad waste of a K-9, but we lost a handler as well. Heat kills. This Airmen literally walked his dog to exhaustion. Seems the poor guy was so scared of Charlie stalking him that he wasn't resting his dog or taking enough watering breaks."

"Your dogs must drink plenty of fluids and have rest or down time, call it what you will. This handler will be sitting in a bunker until he goes home instead of doing the job he was trained for. This could happen to you, so be sharp."

On the posting truck, I listened in on a conversation Vinny was having with Sergeant Lott. Lott was coaching Vinny on how to approach new posts. "If you're lucky to have a washout on your post, use it. Get down below ground level."

Vinny was eager to know more and leaned forward.

"If you have trees or bushes on your post then hiding gets a little easier," Lott said. "Beware of the King Cobra. He hunts at night, it's his hunting ground, and there are lots of rats, big ones."

Not another word was spoken. Even the dogs were sitting still between their handler's legs, tucked under the seat. I disappeared into thoughts of the robot from the TV program *Lost in Space*.

"Danger, danger, danger," said the robot.

I couldn't help but laugh out loud.

"Hey, what's so funny?" said Joe. "A guy lost his dog and his job."

"Yeah," said Vinny. His green eyes looked sad, sort of.

"I think we all feel bad for the handler and his dead K-9," I said.

"I don't think the Airman should have been taken out of K-9," said Joe. "He'll be in a bunker for the rest of his tour."

"We all know this could happen to any of us," said Lott.

Saigon Tea

On my first trip to the nearby city of Qui Nhon, Joe and I met several dozen children begging or selling something. Joe left me wandering the street by myself, but an old timer from the bus ride, age twenty, introduced himself.

"I'm Jerry. If you want to come with me, I'll show you the ways of the city and protect you from yourself."

We left the kids in the street and entered a place Jerry said was the K-9 bar. "All the guys come here," he said.

As my eyes grew accustomed to the low light, I had to wonder how anyone could work in such a place. It was a very large shack with a Coca-Cola sign hanging on the wall next to the juke box. That was it for decorations. I was surrounded by a dozen Vietnamese women in different stages of undress, wearing too much make-up, with half-smoked cigarettes hanging from their mouths as they grabbed for my hands, arms, and crotch.

"Buy Saigon Tea, GI," said a girl in a skimpy yellow dress. Her eyes were painted with the blue of a peacock.

Saigon Tea was a green liquid in a glass the size of a thimble. I wanted a cold beer and the girl wanted the green slime. When I wouldn't buy the green looking crap, then she pressed against me. "You want to fuck, GI? Sucky-fucky, GI."

Not to be outdone, another girl in a scant dress, maybe older, said, "Come on GI, you bad dog handler from the base. We like you, dog man. Come on! Fuck me."

I bought the girl in the yellow dress a Saigon Tea and one for myself, but after smelling mine, I gave it to another girl.

"Dinky Dow GI. You number ten. You no buy Saigon Tea. You number ten GI," said the bar girl. The cigarette hanging from her mouth made her look silly trying to be Western and cool at the same time. I could think of nothing more of a turn off.

I asked where the restroom was. The guys with me just laughed and pointed towards the back door. In the alley, I couldn't find a door leading to a bathroom so I went back and asked one of the barmaids for directions. She grabbed my hand, walked me out the door into the same alley way, and pointed down at a six-inch wide hole in the ground.

"You piss now, GI?"

She made no move to return to the bar but instead she just stood looking at me and at my crotch.

"You piss now, GI, I help you, I help you."

I let her unbutton my pants and pull my pecker out. Without a word, she bent down and gave me my first blowjob ever, in Vietnam. My first blowjob anywhere. It didn't take long, and then another barmaid came out, squatted over the hole, and did her business. She just looked at me, smiling.

I smiled back as I let go in the first bar girl's mouth. She spat it out in the same hole then turned, squatted, and pissed while looking up at me talking about the money I owed her. I paid her.

When I returned to our table Jerry asked how I liked it and how much.

"Ten bucks and worth it," I said.

My friends looked at me as if I was crazy. "You overpaid by five dollars," said Jerry.

"Hey, I don't care if I overpaid. These people are so poor and I have nothing to do with my money except buy sexual pleasures. Give me a break."

Jerry lifted his beer and drank. "Look, this is what you have to learn about doing any business with these people. First off, they think we are all rich. Here you see poverty like nothing back home. Fuck it, that's the way it is."

"Damn right," said the Airman next to Jerry.

"Whether you ask for sex or a beer or a goddamn Bradley Tank," said Jerry, "they will start at double and hope you fall for it. If she says ten, you say five and hold a five-dollar script up to her face, she'll drop her pants if she has any on. If you pay more than the going price, we all have to pay more."

I scratched my head. "Maybe I can start a new trend and have all of us pay double the asking price. Think how that would give a boost to the economy, maybe they could take better care of their children."

Jerry just shook his head and said, "Let's go."

Outside the bar, everything in Qui Nhon stank of piss and unpleasant smelling food. My partner insisted on helping me find my next sexual conquest.

In the street again, I ran into a group of kids who were grabbing at my brand-new watch and trying to rip it off my arm. "Good grief, you little shits, get away from me."

I kept walking and the kids thinned out except for one little boy who was about six or seven. "Hey GI, sucky-fucky. You sucky-fucky. My sister has big tits, you come see, GI. You like, you like!"

"Are you shitting me, kid? Go away," I said.

"GI sucky-fucky. GI sucky-fucky. You come home with me now, sucky-fucky GI?"

I can't believe what I am hearing. This kid is pimping his sister to me.

"GI, come with me now. You sucky-fucky my sister. She has big tits, GI."

He made a gesture of large breasts and I had to laugh. I also had to see his sister, even though I didn't give a shit about large breasts.

"You'll get used to it," Jerry said. He laughed at me. "Isn't it great?"

No, it isn't great.

"Lord, what am I doing? This is crazy, this little kid is out in the streets finding GIs to bring home to screw his sister. He's pimping for his sister. You kidding me?"

"Hey man, go with the flow. Go check it out. At least see if she really does have big boobs. I would be interested in bigger ones. Remember, whatever she asks for, you start at half and work out a price. Have fun!"

Just like that, my friend Jerry was gone and I was following this kid who should be in school instead of pimping for his sister. At his home, he didn't enter the building but left me standing at the door. A small grandmother-type woman met me at the door and beckoned me into a semi-dark room that had a nice smell to it. Must have been incense, but it helped to calm me for some reason. In the center of the floor was a waist high single bed with mosquito netting. The picture of an old man hanging on the wall looked familiar. There were no other furnishings in the room, not even a mirror.

This is a strange whorehouse.

The little old lady asked if I wanted a girl. I nodded "yes."

Grandma disappeared, returning with a young lady who was dressed in an emerald silk robe with white flowers. She was attractive in a farm girl kind of way, and whether she had big breasts was still to be seen because she didn't fill out the close-fitting robe.

"Take off clothes, GI," said Grandma with some insistence.

I ushered her out of the room, through a door with colorful beads and returned to the young lady who had removed her robe and was waiting for me.

I wonder who the guy in the picture is, and why isn't there a chair for me to sit on? Is there a chance of getting a cold beer? And who is that man standing in the doorway staring at me? Is he going to be watching the whole time I'm screwing?

I got undressed and climbed in the bed with her. She pointed at her breasts. "Big tits GI, you like big tits?"

"Sure," I said, and gave her a thumbs up.

From the scuttle-butt I had picked up about Vietnamese women there was no such thing as big breasts in the whole of Vietnam. She indeed had bigger breasts but they had gone flat and hung halfway to the bottom of her ribcage.

She looks young, but could she have a baby sleeping in the other room? Her drooping breasts are a turn off. I can't let her know because I don't want her to feel bad.

I embraced her breasts and showered them with kisses. It was the first time the girl had smiled.

We haven't discussed price. How much are these deflated boobs going to cost me? I'm not going to haggle over the cost for sex. Even though I'll appear 'not too smart' in their eyes, I don't care. I will pay what they ask knowing damn well it will never be enough. If I continue to enjoy the sex, may the pleasure be a win-win for everyone.

Just when I was getting comfortable, Grandma walked in. "GI pay twenty dollar," she said. *What is going on? I'm naked as a jaybird and the door is too far to make a run.*

Grandma left, and I turned my attention back to mounting the sister with the big tits. I had almost forgotten about the man standing in the doorway watching me until I caught another glimpse of him.

Here I am with my ass up in the air like a dog in heat, and a pair of cold, dark eyes full of contempt are watching me. The man's face is familiar for some reason. Is this man VC? Is this whore with the big tits going to be my first and last piece of ass? In Vietnam any sex could be your first and your last.

I was trembling and clumsy, so the girl helped me find Gloryland. After a sweaty climax, we melted like lava lamps into one, in no hurry to disconnect. I gazed into her face and for the first time I saw how sweet her deep brown dewdrop eyes were. With a squeeze, she pushed and giggled. It was over.

Finally, I'm a man in every sense of the word. No more teasing from the other Airmen. I didn't expect her to be so into it. Did she cum, too? Doubt it?

We were lying there stuck together in mutual need when I noticed the shadow of the man watching us again from the doorway behind the beads. *Pervert*, I guessed. Our eyes met and I shot him the finger. He disappeared.

My eyes had adjusted to the light, allowing me to see the picture on the wall was of Ho Chi Minh. But my conquest distracted me with her new wonder. *Baldheaded! Cool!*

"You leave, Airmen, another dogman, leave.

I put my clothes on and paid Grandma for services rendered. Then I handed the girl with the big tits an extra five and headed for the door. She was chasing me while putting her clothes on and begging me to come back. "GI like big tits!" She hollered after me.

I was so glad to be out of that place. I was still looking for a beer or a soda, something from a bottle or can, something cold but no Saigon Tea.

Outside, amid the smells of the street, the man's eyes returned to me. *Could the man standing in the doorway have been her father? Or worse, her husband?*

From the shade of a large palm tree I watched a group of boys coming down the street all dressed in black pajamas. Talking and laughing, they passed me by, not paying any attention to me. They had black cowboy hats on with red bandanas around their necks. They looked like some sort of a gang of Boy Scouts. They disappeared down the road.

The man watching me was someone I've seen before, but where?

I ran into my friend Jerry and together we returned to the K-9 bar hoping to get a cold beer. What I got was a can of soda with a label I could not read but at least it was cold. We were the only GIs in the place. There were probably ten bar girls and one guy behind the bar. He kept pushing the girls over my way to sell me Saigon Tea.

No way am I going to buy any more of that green tasteless whatever in a shot glass. It could just as well be green colored piss. Turning my nose up and shaking my head makes no difference to the girls. After a while I start to feel sorry for them.

Jerry and I never got to enjoy a cold beer; instead we headed to the pickup point for the last bus back to the base.

We noticed an artist by the side of the street doing sketches and stood behind the man, watching him draw. He was good at landscapes of mountains and rice paddies and I asked to see sketches of faces. All the faces were Vietnamese women. I decided to take a chance and handed over my picture of Lillian. He said it would be ready in a week. He was an old man, so I felt it would be okay to leave my one and only picture of her with him.

On the bus back to Phu Cat, Jerry asked about my experience. "How did it go?" Jerry said. "Did she really have nice tits?" He grinned.

"No, they kind of hung down to her waist. She was a good lay but you won't find me in that house again, ever."

"What do you mean?" Jerry looked serious. "What did she do?"

"I couldn't believe that some man was looking at me while I was fucking. It's strange."

"I guess so. Maybe it was another suitor."

"I don't know, what the hell is a suitor?"

"Another paying customer, dipshit. She could have them waiting in line."

We laughed at my ignorance.

"Besides that, there is a picture of Ho Chi Minh on one of the walls."

"No shit?" Jerry said. "I don't think you should be checking out any more of big tits."

"You're damn right. I had Grandma walking in telling me to pay '20 dollar' while I was in the saddle getting ready to plow away. The sex was great, but I'm thankful for not having a bayonet shoved up my rear."

Jerry and I remained quiet as we sped along Highway One, just staring out the window at the country side. The bus crossed over a bridge and I managed to catch a glimpse of a naked woman bathing herself in the river. The scene was sensuous and gave me an uncomfortable erection.

"Damn it to hell," I said. I sat up straight. "I forgot to use a rubber."

"Damn, you are a dipshit," said Jerry. "The clap they have here can get you sent to Japan to die and mamma and daddy get a letter telling them you were killed in combat. The Black Syph makes your pecker rot and fall off. Dumbass, we can't have you missing work and fucking up the off-night rotation."

"Okay, okay, I get it; I hear you loud and clear. Use a rubber. Should I get it checked out?"

"Hell no. What are they going to check out? When you start to drip and the pain sets in you'll know it's time to get it checked out."

We sat in silence for a while longer, then I asked Jerry the question that was on my mind. "What do you think about the baldheaded pussy the women have here?"

"Damn, you must be a mind reader. I like it."

"Are all the girls like that?"

"I guess so, the ones I've been with are baldheaded. I've been told that hairless pussy happens only in the Central Highlands and these women are prized above hairy pussies. Can you dig it?"

I remembered the girl and her six-year-old pimp, and gave him a skeptical look.

I dig it, but Central Highlands, my ass. These girls are too young for whoring.

Chapter Five

Hippie Dog

Phu Cat Air Base, Vietnam – Late January 1969

One of the worst parts about pulling duty on the northside of the base was the jet mechanics working to maintain their planes. The unbearable noise of jet engine run-ups was so intense I couldn't think. I would cover Duke II's ears and wait out the roar of the jet engines at full throttle for as long as ten or fifteen minutes, or so it seemed. We handlers were not given ear protection, but I wasn't about to let the noise deafen my dog.

One night not long after my first trip to Qui Nhon, Vinny and I worked adjacent posts. At the drop-off point, Vinny and I separated and cleared our posts, waiting for oncoming Super Sabres to pass us by, full burners, wheels up, and climbing hard, all this going on less than 50 or 60 yards away and two more F100s right behind. Move? No way. Not until the show was over.

I loved listening to the jets takeoff, but I hated the engine run-ups. They gave cover for sounds of the VC who were always moving up and down the fence lines, through fences, tying off trip flares, tying off concertino wire, making holes so small a 'possum couldn't get through, but the VC could and did. When the jet engines quieted, the VC hunkered down and waited for the next round of ear shattering noise.

Vinny and I welcomed the chance to meet up during the long night because it was the first time we had been together on post. We got around to talking about the thirty days of leave we took before deployment.

"Jimmie, remember the girl I hung out with at Lackland?" Vinny asked. "Well, I went to Mississippi for a visit."

"Cool, man, how did that go for you? Did you get any sugar?"

"Oh yeah, the sugar was good. We had a blast."

"How long did you stay?"

"Got there Friday and left on Monday. She went back to work on Monday and I headed home."

The sky was overcast with no moon that night and there were few stars to give off their light. The air was moving at about ten miles an hour which kept the man-eating mosquitoes at bay. I took the opportunity to unbutton my shirt and let the breeze hit my sweating chest, bringing needed relief from the heat.

"Sharon is a good-looking girl," I said. "Was it worth it to go out of the way to Mississippi?"

"Yes, it was, and that's all I'll say about that." Vinny was smiling, but then his back stiffened. "Do you hear that?"

I heard a tinkling sound in the distance. "What is it?"

Vinny recognized the source of the sound and relaxed. "Oh, it's just some cans tied to the fences or bushes. The VC let the wind do their talking."

"Is Sharon writing you?" I said.

"Oh, yeah."

"Well, I made a detour to north Texas to visit Mary. Same as you, I got there Friday evening and left Sunday. By the time I got home she had fallen in love with some Tech Sergeant and was getting married. It would have been nice to have gotten a few letters from her, but I think the whole Vietnam thing could have spooked her."

"Have you gotten any good mail yet?" Vinny sighed.

"Today I had a surprise at the mail room that made my heart full. My cousin Linda, who lives in Gainesville, sent me a package containing a pair of dark brown wool footies, or whatever they're called."

"That's so corny," said Vinny.

"Hey, she took the time to crochet those for me. Anyway, your sister and Barry's came through with letters."

Vinny's sister went by the name Bunny. She was still a senior in high school as I recalled. I enjoyed letters from Wendy, Barry's sister who lived in Michigan, and I was thinking of making a visit.

"These letters give me something to think about instead of the guilt I feel from losing my girl."

"It wouldn't have made a difference if you could have married Lillian," said Vinny. "Is it as simple as just being her time to go?"

"Lord, I don't know, but I feel like I've let her down."

We separated again, back to our posts until the coffee truck came. I contemplated how each dog had his own personality. If Vinny's dog King's ears perked up, Duke's ears perked up, too. Duke II watched King and did what the other dog did.

Sixty-five dog teams were as different as night and day. One handler had trained his dog to eat a peanut butter cracker by putting it on his nose, and on command from the handler the dog would flip the cracker into the air and catch it on the way down. The dog would hold that cracker on his nose until the Airman commanded "eat."

In the morning Vinny, Joe, Barry, and most the Airmen drank and played card games, as was our routine. These diversions kept us from talking about our work.

It takes strong drink to put a smile on your face. A man plays the hand he is dealt, and I am a man now, aren't I? My choices were a bunker or a dog, so shut up, Jimmie, and do the job you were trained for.

I was eager to find out more about the Koreans and got the opportunity when I saw handlers dressed in karate uniforms going to Camp Thunderbolt. The following day I went with a small group of handlers to watch and see if I wanted to take lessons.

Korea, New Zealand, and Australia provided troops in this war, but being around these Korean Grunts was something special. The Korean Camp building where classes were held also housed thirty or so orphaned Vietnamese boys who had lost all ties to their families because of the war.

The instructor greeted us and explained their program. "The small fee you pay for classes gives us money to run the orphanage," he concluded.

"I'm in," I said. Vinny and Joe were, too.

We bought karate uniforms called a "gi", a set of loose fitting white pants and shirt worn with a white belt and flip flops. We arrived the next day for our first lesson. If we got there a little early, we could watch the orphaned boys drilling tougher training routines than ours. Their discipline didn't come easy for me.

There were twelve students from the Air Force and all of them were from K-9.

Vinny and I were working together and would take the next several weeks learning a routine of different moves.

One day I was practicing kicking on a duffle bag that was full of sand and hard as a rock. I hit the bottom of the bag and jammed my big toes almost to my knee. It hurt like the devil. One of the instructors came over to me, had me sit down, and began pulling and twisting on my big toe. It was all I could do to keep from jumping up and running for my life.

The instructors can't afford to send a dog handler back hurt and unable to go to work tonight or the Air Force will put a stop to us going to karate class. If that happens, the Koreans will lose the money we paid for our instruction and it could shut down the orphanage.

In the second week of February, we tested and received our first belts. Boy, were we proud of our blue belts. A couple weeks later, during one-on-one exercises with a soon-to-be black belt, I took a kick upside my head that nearly knocked me out. I was seeing stars as I staggered home to the barracks. I never went back again.

My Dilemma

January 30, or Tet, the day of the Vietnamese New Year, was two days away and everyone was going about business as usual, except in K-9 and Cobra Flight. The older guys and short-timers were tense because of last year's disastrous Tet.

After almost four weeks on duty working with Duke II, he didn't alert, he didn't take the lead, and he didn't want to walk point.

My dog isn't interested in playing war. This is too scary. I have to report this.

One night, Duke II walked me right off a washout and I fell three or four feet, landing flat on my back. Once on my feet, I realized I had a serious pain in my butt and back. I was mad at my dog for not helping me see through the night.

What? A damn hippie!

I met Sergeant Francis in the morning just as soon as I got off work. I would ask him first and follow the chain of command, so as not to upset some lifer for not following the rules.

"What can be done about Duke?" I said. "Can he be sent back to Lackland for more training, or somewhere in Vietnam, maybe even Japan? What if I'm the problem and not the dog? If I have to take time off from the war and take Duke back to school somewhere, anywhere, that'll be okay with me."

Francis' round face beaded with sweat. He consulted with MSgt Johnson, who decided that Sergeant Francis would run a test exercise that night.

"Sometime past midnight and after the coffee run is over, Sergeant Francis will radio both teams on either side of you and advise them of his coming onto their post and to please not shoot him," said Johnson. "The plan is to see if Duke will alert and whether you can read his alert."

That night I waited at my post several hours after the coffee truck had come and gone. By now it was showing 0120 hours on my trusty Seiko watch I recently purchased at the BX. *Is Sergeant Francis planning to trick me? Maybe even try to catch me sleeping?*

The wind was in our face, there was a quarter moon, and the sky was full of stars. I couldn't have asked for a better night for this test. At 0200 hours on the dot Francis arrived without a call on the radio or anything.

"Damn, Duke, what gives boy?" I whispered.

Duke II should have alerted on Francis well before I could make him out in the dark. I heard him long before I saw him and still my dog isn't paying him any attention until now when we're face to face. If this was VC or NVA, I probably wouldn't be alive.

Sergeant Francis approached me, panting and wheezing. "Anything happen, Moore? Did your dog alert on me?"

"No, Sergeant, he didn't."

Duke II was as glad to see Sergeant Francis as I wasn't glad to see him. His tail was wagging just as hard as it could go.

Is my dog going to give this ole lifer asshole a big sloppy kiss for coming out and visiting us? I'm not happy. My dog is useless as a working attack dog. I have a pet, a mellow hippie dog.

I was near sick to my stomach.

What am I going to do? What are the big shots going to do? I think he already knew Duke was not a military working dog. He knew and was willing to put him out with a FNG of a handler just to see if by some miracle I could work this dog. I can't believe these stupid lifer sons of bitches would set me up like this.

"Airman, are you sure your dog didn't alert on me?" said Francis.

"I heard you and saw you before Duke did. Hell, you were twenty feet away before Duke stopped scratching long enough to pay any attention to you."

Francis just stood there, his face turning redder.

"Sergeant, this sucks. What the hell are we going to do?"

Here it is the twenty-ninth, New Year's Eve and everybody thinking about last year's Tet and I'm out here with a dog that doesn't alert. He'd probably lick the hands of the VC if they offered him a bone.

The Sergeant stared at Duke for the longest time and then looked at me. "You will report to Sergeant Johnson in the morning. I will give him my report and it will be decided then what options we may have and what is to become of Duke."

He has no respect for my dog. I no longer have any respect for my dog. Why should I be the one to have to put my ass on the line every fucking night out here in this shit hole of a country with a damn fun-loving hippie dog?

"That's fine by me Sergeant, but what am I to do in the meantime?"

"Finish the night out and report to Master Sergeant in the morning, Moore," the Sergeant said. Sneering at me like I was dog shit.

"Fuck this shit, Sergeant; I'm out here with a dog that doesn't alert. Would you want to work a dog that didn't alert?"

"Watch your mouth, Airman. You'll report to First Shirt in the morning and that's all I can tell you for now."

"Don't tell me to watch my mouth. You nearly walked over Duke and you're telling me to watch my mouth."

"Airman Moore, we will get you a dog just as soon as one becomes available. We'll talk it over in the morning." With that said, he turned on his heels and left.

Now I understand why the other handlers hated him.

In a few hours, Sergeant Francis and I sat waiting for the First Shirt to come back into his office.

"Good morning, Moore," said Sergeant Johnson. "This won't take long and I'll get you out of here and to bed. Sergeant Francis has made his report and it looks like we are going to have to retire Duke and get you another dog. Just as soon as the next handler is relieved and sent home you will be assigned to that dog."

"Thank you, Sergeant, but what will happen in the meantime? Are you going to send me on duty with a dog that doesn't alert?"

"Airman, you will continue working with Duke until we find another dog."

"Sir, isn't that a little bit dangerous? Sergeant, I got into K-9 to have an edge that would help me get home in one piece and now I've been assigned a dog that doesn't work or alert. That puts my life in danger and the base at risk. Is there anything you can do until I get a new dog?"

Johnson did not answer me. "Sergeant Francis, you're to see to it that whatever post Airman Moore is assigned to, the post on either side will help cover his post also. He will only go out with the early shift until we get this cleared up."

"Thank you, Sergeant," I said. "Can I ask what will happen to Duke?"

"The higher ups will figure that out," said Johnson. "There are several options, but I won't be making that decision. You are not the first to try to work Duke, but we were in hopes you might be able to find some way to get through to him. Until we can relieve you of Duke, I want you to continue working a post each night because you need the experience."

"Is there a chance Duke could be sent home to be retrained or adopted out to some family with kids?"

"Like I said, I won't be making that decision, someone else will. If there are no more questions, Moore, then you are dismissed."

"Sergeant Francis, handlers on either side of Moore's post will cover with him. Is that understood?"

"I'll see to it, sir," said Francis.

I went to the stop to wait for the bus making its rounds. I was hungry for some eggs and pancakes and I was tired, but I planned to drink away my anger before going to bed. *This shit would drive anyone to drink. I wonder how many handlers are coming to work drunk and/or drinking on duty?*

Right after Tet '69 came and went, the monsoon season hit man and dog with a punch to the gut. It rained for 48 hours nonstop and it was hard to imagine it could get any worse, but it kept raining almost every day and night. I switched to the early shift so at least there would be a little light left to see the post before entering the gates of hell, but working in the rain made everything appear as a blur.

One night I met up with one of the second shift handlers who told me his dog was alerting on something, making my heart skip several beats. *Now it's time to walk my post with another handler taking the lead. Humiliating.*

"Could be an animal," I pondered out loud. "You didn't call it in, did you?"

"No! You nuts?" the handler said. It was Sergeant Lott. According to the Code of Handlers, it was against the rules to call in an alert that we could not confirm as human.

"There could be a tiger stalking or a VC in the grass watching our every movement. Who the hell knows? Putting a flare overhead will probably be enough to scare the big cat off, unless the tiger is hungry," said Lott. "But it won't scare the VC hugging the ground."

When a dog alerted and it was called in to CSC, our command center, they would alert the mortar pit who would fire a round and send a flare our way to open up in less than a minute, usually before the exposed dog team could hunker down and hide. When this scenario happened, it left handlers in the open, exposed, lit up like Florida Field for a night game.

"Tell me more." I was eager to learn from his experience.

"Don't fire your flare just because you think you saw or heard something in the bush. The light gives away your position and that of other handlers down the line on either side of you. There better be gooks on your post before you call it in."

"Doing this night after night is turning my brain to Jell-O. I am already drinking a lot."

That night the Rev, working the post on the other side of me, warned me about how rats and rain would set off trip flares and the dog teams would have to investigate. "Stay low, walking up to a burning flare makes for an easy target. There is no other way to check and see if VC, NVA, or a rat had set off the flare but by going and eyeballing it, stay low."

At coffee break, Sergeant Lott put it another way. "Only a 19-year-old kid who is full of cum and dumber than a box of rocks would walk a dog up to a burning flare to see if the enemy has set it off. Get just close enough, find cover, and watch for any movement."

The rain started. Even with my poncho on, by the time I returned to my post I was soaked with water running down my back and overheating in plastic rain gear. My feet, pants, and jacket were dripping wet. I know the temperature was in the nineties, but sweating made me cold. To top it off I had a sinus problem that made my nose bleed.

Poor little ole me.

Duke II was soaked to the bone, but the cold didn't seem to bother him like it bothered me. I had to learn how to walk all over again because the rain made the clay become slicker than whale/fish/dog/snot. After checking and clearing my post it was easy to find the usual spot where dog teams parked themselves for the long nights. Old c-rations cans and boxes littered the grounds. Dumb.

I spotted a sheet of plywood with the usual collection of open cans scattered all around. I was grateful to have a place that was sort of high and dry. It kept me out of the mud, at least. I sat on my bucket with Duke II under my poncho, nothing but his head sticking out.

I heard noises, but I wasn't sure what I was hearing. Even Duke II started to pay attention to the racket when the rain let up and the noise got louder. It was located right underneath me and the piece of plywood I was sitting on.

Rats. I want nothing to do with them. If a rat bites me, there is no telling what diseases I could come down with. Duke, you little chicken shit of a war dog, you want nothing more than to get off that plywood.

It wasn't easy lifting a four by eight wet sheet of plywood that had become stuck to the earth. As soon as the suction turned loose, the board lifted and Duke nearly pulled me down as he tried to make haste away from the area. I held Duke and peered over the edge just in time to see three or four rats the size of small dogs running off.

This is how it ends for me. VC have tunneled in and are using rats to take me out!

I soon came to my senses and found another spot to spend the rest of the night.

The sky cleared and there was a bright half-moon lighting up the early morning sky and me with it, so I could read the letters I picked up that day.

I had written letters to three ladies I knew from high school, and had just gotten my first letter back from a friend of Lillian's. She said she would not be writing anymore because she had a boyfriend and it wouldn't look right.

Hey, it isn't a Dear John letter but it feels like it anyway. I just want a female to care about me, I guess.

Duke II nudged my leg and panted. I poured some water from my canteen into his dish and took a sip for myself.

The next letter was from my parents. They said my aunts and cousin were doing well, made small talk about the weather in Starke, and asked how I was liking the c-rations. I would answer them later.

By the time I met the relief truck, I was soaked and filthy, my poncho the color of clay, and Duke II's fur was wet to the skin. Ten exhausted handlers with ten nasty, wet dogs sat in the endless rain, dogs hidden under their ponchos as the truck flew down the slippery road.

We cleaned our weapons before going back to the barracks. Our CAR-15s were very sensitive to wet and mud. I remembered to bring a towel to dry off Duke II with a good rub down and lots of loving up before putting him to bed. Even though he hadn't earned it I just couldn't help myself.

The only good thing about the rain was getting off early shift and hitting the rack with the rain still coming down. Our barracks had metal roofs which made a pinging sound that would put me to sleep in an instant. It reminded me of rainy nights back home in Starke when I would put Maxwell House coffee cans under the eaves of my bedroom window so the rain coming off the roof made them cans sing.

The fifth of February went by unnoticed but it marked our first thirty days in country and the countdown had begun. Several of the handlers drew up calendars using the word "HOME." Home was divided into the number of months and days left in Vietnam and marked off as the days passed. I bought colored pencils at the BX and used different colors to mark the days.

Isn't it a little strange that I can buy colored pencils where a full-scale war is in progress? Only in Vietnam.

Scared Shitless

I watched as the truck rolled down the berm road until its backup lights came on and it stopped to let off the next team, Vinny and his dog King tonight, and I hoped maybe we would get together. The distance made it impractical.

"Duke, are you as scared as I am? Come to think of it, you've been scared since the day you landed here. For that, I'm sorry. How in the world did you make it out of dog school? Hope I don't trip on any VC. Nah, Duke II would be licking their faces. Well, I guess that would be some kind of an alert."

Shut up. Stop talking to yourself. If I can just get this first part done, I'll find a place to hunker down for the night.

"Useless dog."

I'm working a damn hippie dog who will never alert and here I am stumbling around in the darkness and filthy red clay.

"Duke, good boy, do your thing Duke, you're a good ole boy Duke."

What a load of crap. "Good boy."

We started off again, moving deeper into the unknown ground, stumbling and falling, getting up, continuing forward, always forward to clear my post.

"Duke, did you see that? There are gooks out there in the grass. Do you see them, Duke?"

Please, Lord, let me get this post cleared and find a place to hide. Hear me, God, this is crazy.

My eyes caught a few stars beginning to poke through an overcast sky and I knew for sure it was going to rain sometime soon. The wind came from the east coast and moved the clouds across the sky. It felt good in my face. I could smell the wetness and feel the static in Duke's fur.

Looks like a big storm coming, maybe an all-nighter. What's waiting for me in the swamp tonight?

A picture came into focus; it was the face of my school mate, Jimmy. He was a proud Marine and died that way, just over a year or so ago in this hellhole.

Duke was looking up at me with a strange expression on his face.

"Let's go boy, watch 'em boy, watch 'em." I whispered as much for me as for Duke II.

Does he know? Does he sense my fear? Duke II is looking at me like I'm a fool. A COWARD! Shut up self! Duke still walks behind me. Good grief, I want to go home.

"You chicken shit coward, I hate you, Duke." I heard myself say the words, but I wasn't sure who or what would say such a thing.

I care about Duke II, don't I?

"Fuck you, Duke! If only I had the first Duke, Lackland Duke, now that was a dog. Fuck you, Duke the Second, you chicken shit coward. Fuck you, fuck you, and fuck you again!"

"Jimmie, Jimmie Moore! Who the hell are you talking to, man?" I heard Vinny's voice but I wasn't quite sure what he had said.

I had not moved off the berm road. I was still sitting on my bucket. I had not loaded my weapon or removed Duke II's muzzle. I hadn't checked in with base.

"Jimmie, what's the matter? Are you all right?"

"Yeah, I'm okay. I've got to call in my security check. How late am I?"

"No need to call in, I covered for you," said Vinny. "I walked over to help you clear your post. I waited for you but you didn't show, so I started looking for you. I could see you standing on the road. Then you sat down, not moving. I waited for you to call in but you didn't. I finally called in your radio check and got you set for the night."

"Oh shit, Vinny, you saved my ass. Thank you."

"No biggie man, but Jimmie, what the hell is going on?" Vinny asked.

"I don't know how long I've been sitting there. It couldn't be that long, could it?"

"Twenty minutes, at least," Vinny answered. "Maybe thirty, but you're cool. You were talking ninety miles an hour. Just who in the hell were you talking to?"

"The devil." It was the only thing I could think to say that sounded tough.

"Well, come on, man, we've got to get you on your post. I've cleared it but let's walk it together. I found a little cover for us to hide out until the second shift arrives."

Vinny and his dog King took the point and I brought up the rear. I was talking to myself again and Vinny kept giving me strange looks as we cleared the post, taking our time.

"Chicken shit coward," I whispered in Duke's ear. We stopped to water our dogs.

"Who's a chicken shit coward?" Vinny said.

"This blasted dog, who do you think?"

"Sorry man, I didn't think it bothered you this much. You mean to tell me that Duke doesn't alert at all, nothing?"

"You got it, now? They give me a puppy to do a man's job and sent me out here to get killed. Lifers. I hated them."

"I had no idea it was this bad. No wonder you're talking to yourself."

"Vinny, you remember back at Lackland, why we signed up for dog school, right?"

School flooded back into my mind as I remembered those first days all over again and the anger began to build.

"Vinny, I didn't sign up for this shit."

"Is that why you're talking to yourself?"

"Yes, I guess it is."

"You need to talk to Sergeant Johnson, let him know this shit is freaking you out. You're no coward, Jimmie, you have a fucked-up dog. I don't blame you for being pissed. I would be if I had to walk in your shoes. So tell someone, get it out. How long can you go on talking to yourself and worrying about your next step out here? It's dangerous enough without that kind of shit on your mind."

"I know, I know. But, I can't do it. The lifers will make it tough for me and God only knows how they will take it out on me. People will think I'm a coward regardless of my dog. You know the thinking. You're chicken shit if you fail to work your dog."

"No, Jimmie, I don't agree at all. Nobody is going to make anything out of this; you got the worst pick of the litter. So what are the lifers telling you anyway? What's going to happen now?"

"Just as soon as someone rotates out of here I will get their dog. I still have several weeks walking the dog that doesn't work. Vinny, it scares the living hell out of me. I can't go to anybody about this. I'm supposed to just suck it up and be a man. Lifers."

"Jimmie, if I see you talking to yourself again I'm going straight to the Sergeant and get you relieved from duty until they get you a dog that fuckin' alerts. It is as simple as that. The job sucks enough, fuck Vietnam."

"Thanks buddy, but I'm supposed to get support from the teams on either side of me until I get a new dog. I'll just have to live with it. Thanks, Vinny, for understanding. Let's keep this to ourselves, okay?"

He nodded and I paused.

"Wait a minute, Vinny, I think you may have something there about talking to the lifers. I'm about to go nuts with this dog, I've got to do something.

"I think you should and the sooner the better."

"Okay, what say in the morning that you go to the First Shirt or whoever, even Sergeant Francis will do, and tell them what you saw me doing tonight – all of it. Tell them I will not come in on my own. Hell, tell it any way you want. Vinny, I'm tired of not being able to trust my dog, not knowing if and when something is going to jump up and kill me."

Vinny took a deep breath. "Are you sure about this?"

I was exasperated and unsure of anything. "Oh hell! Let's just forget the whole thing."

"Fuck Vietnam," said Vinny.

The next morning I talked to the First Shirt again. "You'll have a dog just as soon as one comes available," he said.

"What about in the meantime? I'm at risk, the base is at risk, and so are the other dog teams. I can't be depended upon, Sergeant."

"Airman Moore, you will carry on as you currently are doing. The next available dog will happen in less than thirty days. As I told you before, you will work the early shift and the other teams will assist you. Do your job as best you can. Is there anything else, Airman Moore?"

"No, sir."

A few nights later I linked up with Rev again. He carried a *Bible* in his bucket and was reading it when I visited with him. At five feet, he was a whole foot shorter than me. His dog was the biggest in the kennels, and I could tell he was mighty proud to have Rip.

I have a Duke and he has a Rip. Vinny has a King and Joe has a Prince. Who would have thought that such royalty existed in the sticks of Vietnam?

After we had cleared our posts, we stopped and sat on our buckets.

Rev said, "Most times you can read a *Bible* at night just from the light of the moon. When it's crystal clear at night and the wind is blowing I get a sense of something just out of my reach that I just assume is my guardian angel, and I feel so at peace. That's one of the reasons I've been able to stay in Vietnam for so long."

Rev asked me if I knew the verse about putting on the Armor of the Lord.

"No," I said. "The Armor of the Lord sounds like something I should have on."

"You are so very right. Do you mind if I read it to you?"

By the tone of his voice you could tell he was going to read it anyway. Pulling a flashlight from his bucket Rev started reading to me using the red lens so as not to mess up our night vision.

"Jimmie, this is going to be a long read, so settle down, okay?"

"Go ahead, I'm listening."

"The Armor of God." Rev read, "Finally, be strong in the Lord and in the strength of His Might. Put on the full armor of God, so that you will be able to stand firm against the schemes of the devil, for our struggle is not against flesh and blood, but against the rulers, against the powers, against the world forces of this darkness, against the spiritual forces of wickedness in the heavenly places. Therefore, take up the full armor of God, so that you will be able to resist in the evil day, and having done everything, to stand firm. Stand firm, therefore, having girded your loins with truth, and having put on the breastplate of Righteousness, and having shod your feet with the Preparation of the Gospel of Peace, in addition to all, taking up the Shield of Faith with which you will be able to extinguish all the flaming arrows of the evil one. And take The Helmet of Salvation, and of the Spirit, which is the Word of God. With all prayer and petition, pray at all times in the Spirit, and with this in view, be on the alert with all perseverance and petition for all the Saints and pray on my behalf, that utterance may be given to me in the opening of my mouth, to make known with boldness the mystery of the gospel, for which I am an ambassador in chains, that in proclaiming it I may speak boldly, as I ought to speak."

We sat in silence as he offered water to his dog.

"Rev, I have never heard those verses before. I need to write them down."

"You have a *Bible,* don't you?"

"Yes, of course, a *New Testament.*"

"You'll find these verses in the *New Testament*, in Ephesians. You need to give your dog some water. Anytime you are at rest, give your dog water. You can always put it back in your canteen."

"I hear you…Rev I'm not all that familiar with the *Bible* and I just barely can remember the Lord's Prayer."

"Jimmie, I hate being a bearer of bad news, but you got screwed. Forgive my mouth."

He was opening a can of pork and beans.

"You mean when I joined the Air Force?"

"Well, that too, but I'm talking about your dog."

"What do you mean? Look man, I already know about Duke not being a working dog. I'm scheduled to get a new dog as soon as the next one becomes available."

"I'm not talking about that. Everybody who has been here longer than you know all about Duke not being worth a plug nickel. He worked with at least three other handlers before you got him. The lifers were in the hopes that someone would finally click and be able to work Duke. It ain't happened and it never will."

All while he was telling me this, Rev was tearing into his can of pork and beans. He never missed a beat nor took a breath.

"So, you're saying they used me as a guinea pig."

"Yep, sort of." He looked at me with a big moon pie smile on his face. "Jimmie, there isn't a darn thing to be done about it. You can't fight it, there is no one to complain to or even wants to hear about it. But I will tell you one thing, they're not even considering giving Duke any more chances with new handlers to push him off on."

"Well, that's a good thing isn't it?" I let out a long sigh and reached over to give Duke II a scratch behind the ear.

"Of course," Rev said. He drank some water from his canteen. "But they sat by and risked the lives of these other guys who thought they were given a good working dog. And people wonder why there is so much animosity for these sorry excuses of men who make careers out of covering their tail ends and not the troops."

"I can't imagine spending a lifetime in the service with the only goal being to retire with the highest rank possible."

"Don't trust anything a lifer tells you. They'll have you back on second shift before you know it."

Chapter Six

Under Attack

Phu Cat Air Base, Vietnam – mid-February 1969

US Headline News

Final Issue of *The Saturday Evening Post* Hits Magazine Stands.

New Boeing Model 747 Makes Maiden Flight.

Every morning before going to bed, Airman Bolden would listen to his record player playing "Little Green Apples." As the card game broke up and with the rain pouring down, artists like O.C. Smith and Al Green playing their soft soul music put a lot of us to sleep.

Bolden said he had grown up in a bad part of Miami, though I thought all parts of Miami were bad. He was a family man, with pictures of them everywhere highlighting his handsome, olive brown skin with eyes to match, alongside his beautiful wife and daughter. He stood about five feet, five inches, smoked, enjoyed a drink with the best of us, and was a very good card player who won a lot.

Bolden's wife was soon to give birth and we were happy for him, so we needled him. "Mark, you hear from home? How is Momma doing?" We had a bet going on whether the baby would be a boy or a girl.

Bolden wanted a boy this time because "the house was full of women." His words.

The morning of Valentine's Day, Jim, one of the seasoned guys in our barracks, had overheard my horror story from my last trip to town. "Why don't you head to Qui Nhon with me today? We'll hit the K-9 bar and grab a beer and some girls," he said.

It was Valentine's Day, the fourteenth of February. *Ten months to go and counting. I'll take my mind off hippie Duke, work, and the rest of this shit by going to town and getting laid. I wanted a cold beer and nookie in that order.*

"Sure, I'll go, but what's so important about going to the K-9 bar?"

"I've had a lot of fun there over the months and gotten to know the girls, well, very well."

Jim was tall, broad shouldered, with solid muscles, intense green eyes, and light brown hair. Back home in New Hampshire, he had been the star quarterback on his high school football team and the Homecoming King as well. I got an earful.

Once in Qui Nhon, I marveled at the way the girls flocked to Jim when we entered the K-9 Bar. I found two lovelies, one for each arm, and nobody was grabbing at my crotch or asking me for a Saigon Tea this time. They found a cold beer for both of us.

The two ladies trying to sit on my lap gave me an aching hard-on. I wanted to take both of them somewhere to get some relief. "Let's get out of here. I need a woman." I said and headed for the door.

"Okay, okay, I'm coming," said Jim. He untangled himself from a dozen lady claws trying their best to hold him back.

"Okay, girls, gotta' go. I'll be back."

"You number ten," she said. She was teasing now. "Why you leave me? I no want Saigon Tea. Come on, GI Jim, Please! I number one sucky."

I had to laugh at the scene. Girls were exposing themselves as he made tracks for the door.

"Run for it," he said.

After a block we slowed to a walk and caught our breath. In the street, I could see the people hated us and our conquering hero attitudes. No one on our side gave a shit.

I noticed half a dozen teenaged boys dressed in cowboy hats and wearing red bandanas around their necks. *Oh, those guys again.* They all wore G.I. combat boots and had brass buckles holding up their Levi's jeans. The leader of the gang wore a white cowboy shirt with snap down buttons, while the rest of the boys wore t-shirts with Coca-Cola and Miller Beer printed on them.

Were these the same bunch of kids I saw on my first visit? "Jim, what the hell is that?"

"Oh, them? They call themselves The Cowboys. They're harmless."

I almost wanted to laugh but held back. It wouldn't be cool to piss off this bunch of John Wayne imitators. As they passed us by, they hissed unrecognizable words and stared, not hiding their contempt. *Yep, we're glad to be here, too.*

"Nothing to worry about," Jim said. "Just Vietnam's version of a New York gang."

"They may look like a bunch of idiots, but we're killing boys that look like them all over Vietnam, and you say they are harmless." *I should have stayed on the base.*

"Calm down, Jimmie. I've been coming here for ten months and never had a run-in with them. Would I bullshit you? *Probably.* Hell no, so fuckin' lighten up. Come on, its time to get you laid. You're makin' me crazy."

The urgent feeling to get laid was long gone.

We left the main street, a paved road and one of few in Qui Nhon, and entered the back alley ways that seemed cold with dampness. It didn't make any sense at all, not in a hundred plus degrees of heat. The stink of piss mixed with the smell of boiling fish heads, dog, and who knows? The offensive odors made me want to pack it in and head for the bus.

"Jim, how does anybody fuck with this stinking smell?"

Jim stopped and turned to me. "In Europe, the French developed perfume to smell good and to cover body odor. Over here, the Chinese created incense to cover up all the other smells. Just wait, you'll see."

We entered a house with no windows. The room was lit by candles. As muggy as it was on the outside, the inside was cool, and the smell took me back to the first time I smoked grass in Gainesville.

Jim had been in his element ever since we got off the bus and it was no different here. The house's main room held a comfortable sofa under the picture of Jimi Hendrix, and there was a bathroom that had a primitive rope-and-bucket shower with a pee hole on the floor. There were two small refrigerators stacked on top of each other, and Jim pulled two beers out, handing me one as he headed for the sofa.

A woman entered the room and smiled at Jim. She walked over and sat down on Jim's lap. She was attractive, and I knew this was no bar fly. She wore the customary black silk pants, and a silk top of white with lots of lace at the collar.

She looks so out of place and so very beautiful. I can feel the blood beginning to move to the center of my body. Damn, I want this woman.

"This is Sue" said Jim.

Her English was better than the girls at the bar, or at least I could understand her a little better. Sue brushed Jim's ears with her small, painted mouth and whispered something that got a big laugh from him. They both looked at me where I was still standing and laughed again.

Sue said something in Vietnamese. Another woman came from the same doorway and stood looking at me. I could see she was nervous, but I could also see that she looked like Sue.

Maybe she is Sue's younger sister? The candlelight does much to favor her natural beauty.

"Jim, how much is this beauty going to cost me?"

I must have had a silly grin on my face because all Jim said was, "It's on me, Tiger."

Sue looked at me and said, "GI looks hungry, Jim. You tell him to go easy on my girl, all right? You tell him to go easy."

Jim shrugged his shoulders. "Take it easy on this one, okay?"

"Sure," I said. "Of course."

I never got the name of this girl and she was dumber than a rock when it came to speaking English so I just let it go. She took me into another room with a small double bed and a nightstand with several candles giving off their soft light.

I'm not nervous this time. In fact, I feel quite at home. It's amazing what a few candles, incense, and a cool beer can do to set the tone.

The girl unfastened my belt and motioned me to undress, while she pulled her t-shirt over her head, exposing dark nipples on tiny brown breasts. I stood naked in front of her as she sat on the bed. She began to stroke me and I succumbed to her gentleness. I sat down next to her and for the first time we looked at each other.

Something's not right about this.

She pushed me to lie back, and at the same time took me in her mouth. *Damn, that feels good.*

Then I felt her teeth bite me and I jumped up with a squeal, pushing her away. Eyes wide, she put one hand over her little mouth and giggled. *Maybe she doesn't realize she just took a nip out of my pecker. It doesn't hurt, it's just the shock of being bitten.*

She went back to her task and I began to relax by taking a long drink on my not-so-cold beer.

"Damn," I yelled. This time her nip drew blood, and it hurt. "What the fuck are you doing?"

I had to see if there was any damage, so I got up to check. *Is this little girl a VC*?

When I reached for my clothes, the girl went nuts on me. She was almost in tears as she begged me to lie down, but I'd had enough and was ready to get the hell out of there.

She started removing her pants, then it hit me again that something was not right. *No hair. Shit, who cares? She's a fuckin' hooker.*

She lay back and spread her legs. "Fuck me, GI, fuck me."

I couldn't. In the soft candlelight, I saw her pussy had only the faintest hint of soft peach fuzz, and it stopped me cold in my tracks. I began putting my pants on. The little girl burst into tears.

"GI number one, number one GI. Fuck me, fuck me, GI."

By the time I was almost dressed, Jim and Sue came into the room, both pulling on their clothes. "What's going on Jimmie? What the fuck happened?" said Jim.

"She bit me and I'm bleeding," I said. I felt embarrassed as everyone was looking at me.

"No shit! She bit you? You're joking. Let me see it."

"No, it's okay; I just want to get out of here, now."

"Let's see it, drop your pants. Mommy got to see this."

"Did you say Mommy, as in Mother?"

No one answered me.

I dropped my pants for all to see. Sue came close for a good look at the now drying blood and gave me a look of pure horror. She turned and slapped the girl, nearly knocking her off the bed. Sue then began removing her own clothes, at the same time pulling down my pants.

"Sucky fucky GI, I make right?"

"Two for one. Not bad at all." I could hear Jim from the doorway.

"Jim, this girl is underage."

"Is that what this is all about? Young pussy? Are you shittin' me? Jimmie, you fuck both these bitches. It's a matter of life or death, they need the money."

"I don't give a shit about the money but I will not fuck this little girl, a baby girl at that. Fuck no, I won't."

I had two women, or one woman and a girl-child, on their backs with legs spread. Both were screaming at me. "Fuck me, GI."

I threw a twenty dollar note on the bed and ran for the door, pulling my shirt on as I left. Again, I heard the slap land against the young girl's face and knew that one of us had done something very wrong.

I didn't stop running until I reached the main street. I patted myself on the back for having the good sense to remember how I got to the house and back out again. The running had opened my sweat glands and a small breeze offered some comfort as I sat down on the curb for a short breather. I was hoping that Jim would be close behind me and so I waited.

My eyes refocused to my present condition, and I noticed the artist who was to make a portrait of Lillian was sitting at his little work bench just across the street. I planned to hang it over my bed so I could see it before going to sleep. I approached the little man, and he looked up from his work with a slight smile that disappeared.

There are few smiles to be seen outside of the Air Base. Everything looks grim, downright ugly, and the people don't smile. Americans are always smiling. If by smiling we could win the war, this one should have been over before it started.

The old man reached into a bag and pulled out a drawing in colored chalk. I was disappointed, but I just smiled the same ole stupid American smile. "How much?" I said.

"Ten dollar," said the old man. He searched my face for objections, but I gave him none.

He had difficulties with Lillian's western eyes, but it's not so bad. I'll keep the original close to me and find a safe place to dispose of this one.

I paid and returned to the K-9 bar with my portrait, still searching for Jim. Several other men were in the place so the women left me alone. I ordered a beer and got a cold one, so I paid a nice tip for it.

I sat alone drinking my beer and looking at the real picture of Lillian, comparing it with the chalk picture. In the artist's rendition, my Lillian's eyes were slanted, and the colored chalk looked like the weird colored pictures of Elvis on beach towels at the flea market.

Oh, my dirty heart. Here I am thinking about getting laid on the very same day I picked up Lillian's finished portrait. It isn't right.

At last, Jim arrived. A kid followed him in, carrying a box. As soon as Jim walked over and sat down at the bar, the kid asked if we wanted our boots shined. We didn't. I handed the boy a couple bills of script which he accepted without a "thank you." He approached several other tables before one of the bar girls asked him to leave and escorted him to the door, the kid cussing her for everything he was worth.

It was time to get back to Phu Cat. Jim and I made our way for the bus stop.

The sun, the sweat, the breeze all combined to put me in a melancholy state of mind and Lillian's face came to me. *I remember a chilly fall day at a friend's house holding Lillian in my arms as we lay on a couch. It felt like love and it was warm-pleasant and unforgettable.*

On the ride back to base, I shared my observations with Jim. "They never really smile, but the whores at the K-9 Bar offer us laughter and forced smiles that look like panic."

Jim shook his head at my naivete.

"When the base was built, thousands of families were displaced from their farms and moved into the cities," said Jim. "The homeless families, along with everything they owned, were just picked up one day and dumped into the city of Qui Nhon."

My eyes must have been bugging out of my head, but Jim paid no mind. "Uncle Sam promised these displaced men and women jobs working on the new Air Base, but…"

I interrupted. "I can imagine no matter how big a base is, it will not be able to hire everyone."

"Right, Jimmie, and the ones who were lucky and got a job are resented by their neighbors."

"So, they were forced into the slums of Qui Nhon, and left to making a living any way they could? No wonder there's so much stealing and prostitution."

Fear of going hungry, or a grenade being thrown through the door as you sit on a GI's lap. There is nothing to smile about.

Jim finished his story and I thanked him for the tour. We parted our ways with him going to the chow hall and me headed for my quarters.

The barracks were in an uproar when it got around that I had just returned from town. "Were you there, Jimmie? Did you see it?"

"See what?" All I wanted was to find my bed and crash for a while.

"Didn't you hear the news from Qui Nhon? Our K-9 bar was blown up."

Other than me being in the place just minutes before the bomb went off, I couldn't tell them anything. The rumors said several of the bar girls were killed and the place was leveled.

Damn! I still have over ten months to serve in this shithole.

After supper, I picked up some mail and was eager to catch up on the news from back home. My parents had written a short letter letting me know that all was well at home. Mommy did most of the writing with Daddy writing just a few lines to say, "Hurry home."

That night, at guard mount, Sergeant Francis offered a few more details about the K-9 Bar that was blown to hell and back. "VC did it," he said.

Like, no shit Sherlock; tell us something we don't know.

"Turns out a small boy of about four-to-six years old went into the bar and left a shoeshine kit containing explosives. It was probably on a timer, since there was no small body found in the destroyed building."

I gulped.

"Three bar girls and the owner were killed outright with five more women wounded," said Francis. "No Americans or Allies were killed."

That little boy I saw in the bar was responsible for all the mayhem. There are just too many ways to die over here.

Sergeant Francis finished his update with a new edict. "As a result of the bombing, the City of Qui Nhon is officially off limits to all Air Force personnel."

I was glad, but Joe's eyes filled with tears. I suspected he had a girlfriend in town. Most of us did not understand why anyone would risk jail time for disobeying a direct order putting the town off limits.

Joe had good reasons for risking his life and hitch-hiking the twenty-five miles. He was in love, again. He would just up and disappear without telling a soul where he went. Sometimes Vinny and I would go check out our usual spots looking for Joe. After awhile it became obvious to us that Joe was sneaking off base and hitch-hiking to Qui Nhon to see a girl. But in real life, the damn VC said to cut the shit out, leave our women alone or else.

It took Joe less than thirty days to find a college student named Vera, the love of his life, in war-torn Vietnam. At five foot two she was taller than the average Vietnamese woman. The nice thing was she didn't ask for money at their first meeting when Joe first met her at the K-9 Bar.

In the bars, the women's jobs were to attach themselves to any serviceman who came through the door. Instead, Vera invited Joe to her own place, gave him an ice-cold beer and a hot bowl of rice and cabbage. They made love and drank beer and made love again until it was late and Joe had to catch a bus back to Phu Cat. She walked him to the bus stop but would not kiss him goodbye, because it could get her killed.

Joe confided in me that he felt bad at first for not offering anything more than a heartfelt thank you. Almost seven months later he still risked jail time, and life and limb, to visit this Vietnamese woman. She always invited him back and the thought of sex with her again made him think next time he would bring some flowers, maybe some candy from the Base Exchange. Still, he sometimes wondered if he was being set up.

This is Not a Drill

It was February 23. A loud voice woke me from deep slumber. "The base is under attack. Get your gear and meet at the pickup point immediately."

It was Master Sergeant Johnson's booming baritone I heard. "Move it men! Got to get posted. Thank God there is some light left."

I shot out of bed. Men were scrambling all over the barracks.

"Shit," said Joe.

"The enemy's hitting us in the twilight zone," said Vinny.

"I thought the VC hit early in the morning. What's going on?" It was Barry, still waking up.

More yelling burst from outside the barracks. "Move it men, let's go. Shit."

Inside, "Little Green Apples" was playing in the background. I dressed, jumped into my boots, zipped them up, and was ready to go.

Thirty of us jammed into two vehicles and we hung on for dear life as the driver flew down the paved road that lead past the bomb dump and the fuel storage area on the way to the kennels. On the tense ride, I brooded. The Men wilted. Yes, tonight! We were Men.

The enemy will be close by, maybe even hiding in the bushes and ditches along the way. An enemy could dispose of the only threat to him by ambushing the truck and killing every handler on board. There are only two weapons, one in each truck, to defend us if we are attacked by a determined enemy. Only in the fucking Air Force.

Ten minutes after the first warning back at the barracks, the second shift stood at attention for guard mount. Staff Sergeant Francis was barking at us all.

"Kilo 9 has sappers inside the perimeter. Get your gear loaded up as quickly as you can. We have a handler down but no other information."

Airmen all around me were wide-eyed and the dogs in their kennels were barking like crazy.

"Stay off the damn radios unless you have an alert and have verified the source. Don't waste anybody's time on false alerts. Any questions? Okay, move out."

We retrieved our dogs, loaded up, and traveled in reverse order, dropping off handlers and dogs. When I dismounted with Duke, the sun had just set and there was enough light to see the terrain of my post.

I hate this dog. He could get me killed. My only hope is for a friendly handler who will be able to help me clear my post. At least I know the post and where the washouts are good for hiding.

Once I was locked and loaded and had Duke in his working collar, we headed onto Kilo 23, pronto. On my post, I found the added benefit of a bunker manned by three security policemen. I introduced myself and asked if they had heard anything, but it turned out they knew less than I did.

"I will be hiding in those bushes. Please don't shoot me," I said. They took the ribbing with good nature.

In a bunker, you are surrounded with sandbags that offer some protection. Duke and I can hide behind my bucket. Ha Ha! I laughed to myself at my joke, but it wasn't funny.

Two handlers helped me clear the post. Then I made ready for the longest night of my life. I could hear my own words as I prayed, "Oh God, protect me and Duke tonight and let us live to see the light of the sun."

The shadows of flickering light from the many flares in the sky reminded me of the movie *Fantasia*, where the brooms with buckets carrying water cast shadows on the wall while they just kept coming and coming. In the falling night on Kilo 23, the whole gook army could disappear.

"Kilo 23, radio check."

"We read you loud and clear Kilo 23."

"Copy," I said.

Hell, let's act like nothing is happening, and please keep me in the dark, by all means. Not even a "Fuck you very much" or a "Keep your powder dry and your head down."

I was just off the berm road when I spotted Rev approaching me from his post.

"Boy, am I glad to see you," I said, "I just radioed in and not a word about what is going on."

"That sounds about right; they are pretty busy about now and no time to be worrying about us. We'll be all right. We pulled the lucky draw, and the Lord is with us."

"What do you mean, we pulled the lucky draw?"

"I mean you are on the right side of the base and out of harm's way," Rev said. "This base has never been attacked other than the northeast side where the bomb and fuel storage facilities are located. That's Charlie's M.O., so I feel comfortable telling you that the action going on right now is limited to that side of the base. "

"So it's business as usual for the rest of us."

"That's right, doesn't mean we are not extra careful, but it does take a little of the pressure and worry out of the job. I'll get back to my post after we've walked your post together and see what we got."

I was glad to have Rev with me as Duke and I began walking Kilo 23 with Rev and his dog Rip following close behind. I still was stumbling around in the dark, but managed to stay on my feet. By the time we finished clearing Kilo 23, we were soaked to the bone with sweat, I needed a break, and looked like Duke needed water as well. The last words from Rev as we went our separate ways were, "...to water my dog. Back in an hour or so."

Sergeant Rev will make a great Preacher. I will miss that.

For my part, I found the washout and made myself comfortable for the rest of the night. Next order of business was to get some water for Duke and let him rest. I needed water and rest myself, but the dog came first.

Sergeant Collins and I later met up together near the bunker that was on his post next to the berm road. We wanted some news, anything to let us know what was going on. The three men in the bunker didn't know any more than we did. The night sky was lit up over the far side of the base and it had been so since darkness fell. There were planes circling and a single chopper flew low to the ground as it made its way parallel to the berm road and out in front of us. Some mighty big guns were in the night sky and they gave us a sense of security. I almost wanted to let out a cheer as the chopper flew by, but I would do that when no one was looking.

After walking almost every inch of Kilo 23, looking for an enemy I didn't want to find, I returned to my little washout, put some water down for Duke, and opened a can of beans and franks, or beanie-weenies as they were called by the troops. I stared straight into the darkness and the night stared back at me. I couldn't see a thing.

I can't be mad at Duke, but I feel damn naked out here. What will happen to this dog that I've reluctantly grown to love? He needs to be retired from the Air Force and put up for adoption.

The north side was getting hit hard, or it appeared so from all the flares dotting the sky. From this distance, we couldn't hear gunfire. There was nothing on the radios to let us know that a fight was in progress or that it was over already except for the cleanup. We were too far away to hear any weapons firing, but overhead I could see the red flashing lights of several aircraft flying circles around the area of the base under attack. Command was taking no chances.

It will be a long and lonely night on Kilo 23, but I'm grateful not to be located anywhere near the fuel and bomb dumps.

I lifted my face to the stars and the moonless sky that had been lit up all night with a hundred flares by now. The aircraft flying above was the only remaining sound of the fight that took place tonight. Around 0330 hours the moon had shown its welcomed face and everybody took a deep breath. The higher into the night sky the yellow half-ball moved, the brighter it got.

My head was the only thing visible above the rim of the washout. I had the perfect hiding spot and all the light I needed to disappear somewhere. Duke checked for snakes. That was one job he could do, at least.

What about the guys in the sixty-foot towers? I don't want anything to do with that. Talk about hanging your ass out in the air.

We were to keep all weapons on "safe" as the waiting continued. The night sky was lit up with flares and it had been so for the past seven hours. There was no contact with the handler on Kilo 9.

Later, I took another hike up near the berm road and walked upon a mumbling handler and his dog.

"Hey, Vinny, who you talking to, man?" I said.

"Damn radio! They can't be sending us out here without being able to communicate. This fucking sucks."

Vinny was talking to himself, but Sentry was eyeballing him like they were having a conversation. Vinny picked up on Sentry staring at him and continued his self-abuse out loud.

"You know Sentry, the base is on full alert status, the fucking VC are inside the perimeter, shots are being fired, people are probably dying, and I can't even report in or hear a damn thing on this damn radio. This sucks, boy, this really sucks."

Sentry was locked on every word being said, and Vinny had to laugh and give Sentry a big loving up. Vinny found much comfort in having this fearless dog and being able to communicate with someone.

"Shit, Sentry," he said, "all I want to know is if it is possible other posts were attacked." The sun was starting to peek through the darkness and there would be no more attacks tonight.

"Vinny, what's going on? Does your dog talk back?"

Vinny's green eyes were cast with fear. "What the hell is happening, man? I've been without a radio all night and I don't know whether to be scared shitless or get ready for pickup. What in hell's name is going on down there?"

"Vinny, believe it or not, I can't tell you anything because none of us know shit. I have seen nobody but the Rev since we were dropped off."

"Well, I'm sure glad you decided to walk down and check on me, it's been a horrible night. Does Base know whether I'm dead or alive? Does Base give a shit? Damn, Jimmie, I'm pissed."

"Yeah, base has tried all night to get ahold of you, so your radio is dead. Think maybe we need new radios or old ones that work, something. Can't send smoke signals at night."

Vinny didn't laugh at my little joke and I didn't blame him one bit.

"Anyway, Vinny, I better get back. I see several headlights headed this way. Get back down on your post – you're about to be rescued, I guess. Maybe relief is coming. See ya."

With that we moved out and I returned to my post, four or so blocks down the road.

I'm going to die in Vietnam.

Good News, Bad News

It was close to 1000 hours before the relief truck started picking up the dog-tired dog teams. The sun was already blazing. The dogs were taking their lead from the excited handlers and were up and moving around the bed of the truck. It wouldn't be long before there would be a fight. A dog fight.

I was amazed to hear that four Vietcong had been KIA on Kilo 9's post. One gook had been wounded and captured, and since Phu Cat Air Base had no jail or stockade or whatever the Air Force called it, the rumor was that the captured man would be imprisoned in one of our kennels.

I found a seat and slid Duke between my legs. "If they put a prisoner at the kennels, the dogs will never get any rest," I said.

"Fuck him, anyway," said a giddy Airman as he stood. "I volunteer to let them put him in with Trojan, just for one night."

That got a big shout of laughter and the dogs started barking and trying to attack each other. Our overloaded truck bounced over the clay roads that hadn't been plowed from the last rain we received.

The truck was a madhouse of barking, fighting dogs making a worse racket than the NCO club full of drunks.

Handlers added to the chaos by dressing down anybody who happened to be walking down the same road we were on. We cussed officers, Security Police in their little cute blue helmets, even mama-sans on their way to work. Nobody was immune.

An excited Airman yelled at an NCO who was just trying to get to work. “Hey Sarge, did you sleep well last night? Well, fuck you very much.”

Some of the guys were standing and screaming at Airmen who hadn’t done a thing except to be walking on the same road to work. “How did you sleep last night, dipshit?”

We got hit by the enemy last night and lived to tell about it, so we cussed them all. K-9 would be back at it, tonight.

No matter what we said, they either didn’t react, or they just waved or gave us a thumbs up. It sunk in that our comrades on the base appreciated our efforts and that the base was secured. They were just as happy to be alive as we were.

At the kennel drop-off, most of the dogs were dragging with tails between their legs and the ears not in the usual pointed position. First thing off the truck we made sure our weapons were cleared, no ammo left in the chamber. Once that was done we took our dogs to their individual kennels. Duke II and I took a detour to relieve ourselves.

How many dogs will get sick from the long night of duty? How many won’t be able to go on duty tonight? When we’re shorthanded it means we have to double up and fill in larger gaps in the defensive perimeter.

“Airman Moore, report to the First Shirt before heading to the barracks. Got some good news for you.”

I didn’t see who said it but I got the message.

“Okay, I’m almost done here. Thanks.”

I finished putting away my radio and weapon and walked the short distance to the First Shirt’s office. I knocked on the open door, poked my head in, and was waved in by the Sergeant.

“You wanted to see me, Sergeant Johnson,” I said. I walked in, and came to an uncomfortable attention before his desk.

“Have a seat, Moore. I’ve got good news and bad news for you, which do you want first?”

Are you shittin' me? The good news, bad news, and make your choice routine. This guy watches TV too much.

"Well Sergeant, if the bad news is about Duke then let me have that first."

"No, Airman, I'll tell you the good news first simply because you deserve good news after working a dog that needed to be retired a long time ago."

Then why the hell did you bother to ask me?

"I appreciate your toughing-it-out attitude and going about your duties, giving us the time to work out the problem as best we could. The good news, Airman, is as of last night with the loss of Airman Lane, you will be taking over his dog immediately."

"I'm not sure I know Airman Lane, sir."

"He's assigned to the other barracks. You know, that black guy from Chicago."

"Oh, yeah. He's a pretty cool guy. Is he going to be okay?"

Johnson ignored my question. "The bad news is you are going TDY to Saigon to have Duke put down."

"Put down, sir?" The words caught in my throat. "You mean they are going to kill this dog who could be retrained or returned to civilian life? Sir, can't they put him up for adoption? What a waste, what a damn waste."

"Airman Moore, you will get a hold of yourself, right now."

"But sir, is there no other way?"

"I've had an eye on Duke for almost a year. He's had three other handlers who had no better success working that dog than you, so get over it, Moore. That dog has put you and other handlers in danger and it has to stop."

"But –"

The impatient Master Sergeant interrupted me with a raised voice. "Enough, Airman." I flinched, but I didn't think he saw it.

"Moore, we've had a rough night. We've lost our handler Marcus Lane to an injury and we are all very tired. Here are your orders to take Duke to Saigon and to turn him over to the Veterinarian by the name of Captain Rogers. He is expecting you. Your flight will leave at 1420."

"Yes, Sergeant, I'm sorry, Sergeant. I – I know I shouldn't be emotional. Sir, what is my new dog's name and where is he located?"

"Moore, it's okay. When you get back and things settle down here we'll have another talk. You'll find Marcus' dog Junior in kennel 21. His tattoo number is X480. He's yours now. Get out of here, try and get some rest. Don't miss that flight or I'll have your ass."

"Yes, Sergeant," was all that was left to say. I turned to leave the office.

"You are getting a good dog to handle, so take some comfort in that. Now, get out of here."

I was sad and happy all at the same time. I didn't know whether to go see Duke and cry a little, or go meet Junior, introduce myself and let out a rebel yell at my good fortune. I decided to see this new dog of mine.

Junior lay in a corner, curled up in a tight ball with his eyes closed. I knelt down and without making a noise I stared at my new dog. After a few minutes, Junior opened his eyes but didn't lift his head. He just stared back at me.

"Junior, good boy," was all I could think to say. I didn't want to disturb him and get him barking at me.

"Junior, you're a good boy. You've had a rough night, haven't you boy? I know you're tired. Good boy, Junior."

His water dish didn't need refilling but I decided to refresh it as a way of introducing myself and letting him know I would be taking care of him now. I got the hose and, with as little splash as possible, I ran a little water into his dish. He never lifted his head as I set down another dish of rations. I sneaked a can of liver, added it to his dog food, and set it inside his kennel. He didn't move. He just stared at me with those sleepy eyes. I decided to leave him alone.

"Junior boy, I know you miss your Marcus. He is going to be alright, boy. Junior, Marcus is okay and you are one brave dog."

Marcus' Story

I left Junior and caught one of the trucks taking the rest of the guys back to the barracks.

"Moore, you really got yourself a good dog. I felt bad for you having to work Duke but that is behind you now. Good luck, Jimmie."

"Thanks, Lott, I appreciate that."

Sergeant Lott was sitting next to me in the truck bed. He was a short timer, due to rotate back to the world in less than thirty days.

"I didn't know Marcus very well. Has anybody heard what is to become of him?" I said.

"I was on Kilo 11," said Lott. "I heard Marcus was wounded and running out of ammo while all the shots were being fired."

"I couldn't hear any weapons from Kilo 23," I said.

"When the gunfire stopped, base radioed to Kilo 10 that we had a man down and to go check on him. But no one knew for sure if the area was secured. The Vietcong could be out there and waiting for us to make a move."

When Lott paused, Barry Teagan chimed in. "I was on Kilo 7, but I was nearing Kilo 8 when I saw Sergeant Francis, along with two handlers acting as backup. He was driving the deuce-and-a-half over the side of the berm road, not knowing whether there was still enemy firing or not."

Barry continued to describe what he had gathered from the nearby handlers and the nervous guys in their bunker. He said Francis flew his truck toward Marcus' bucket, stopping when he caught the dog standing in his headlights. On the ground lay an unconscious handler. As the three men walked up to the wounded Marcus, his dog decided to lay down across the Airman's legs. Sergeant ordered one of the Airmen to put a muzzle on Junior.

The young Airman approached his wounded comrade, but Junior showed his teeth and was growling. By this time there were other security forces coming on the post in an attempt to help with the rescue. This made it worse and Junior became even more agitated and aggressive. The wounded handler was still on the ground with a Sentry Dog protecting him while he was bleeding to death.

Sergeant Francis wrapped a jacket around his arm and approached the dog. "It's okay, Junior, easy boy, out, Junior, out".

Junior was still growling, but he got up and off his wounded partner and gave way to the muzzle.

Sergeant Francis patted the dog and praised him. "Good boy, Junior, you're a damn good dog. Easy boy, easy."

Junior calmed down but never took his eyes off his master, who lay motionless. He tried to move forward and protect the handler as the medics were working on him. The medics administered morphine. Marcus was still bleeding from a bullet wound that entered at his shoulder and came out at the elbow, and was receiving plasma as they moved out. The medic said the Airman would live but the arm was an unknown.

I was impressed with Barry's account. "He's headed for the world with his million-dollar wound."

"Yes," said Lott. "A bad wound to the arm and shoulder. But I did hear he is on his way to Japan, and that tells you it's a damn bad wound. You wouldn't have gotten Junior assigned to you if there was any chance of him returning."

"I hear ya, man. I know that." I smiled from ear to ear because a huge load was lifted from my weary shoulders.

Back at the barracks, TVs were blaring because we were hoping there would be some news of the attack at Phu Cat. Radios were blasting for the same reasons. Several tape players played continuous music without commercial interruption, and over it all I could hear the song, "Little Green Apples."

The poker game was five card stud with deuces wild. The beer was cold, the bourbon warm, and there were plenty of smiles all around the table. I drank several beers and I was feeling good.

I won't be getting any rest before meeting my plane. No one has any interest in going to bed, except maybe the Rev.

Chapter Seven

A Chance to Say Goodbye

Phu Cat Air Base, Vietnam – Late February 1969

The several beers I had consumed that morning did nothing to calm my high octane reaction to the past night's fighting. People had died and one of our own lay in a hospital bed with a severe arm and shoulder wound.

There's talk going around that he could lose that arm, and rumor has it he won his ticket home with that wound. I sure hope he makes it okay.

On top of that, my new war dog was a war hero. Junior, as in Junior Thornton, got it.

When 1400 hours neared, I picked up my TDY (Temporary Duty assignment) orders for Tan Son Nhut Air Base in Saigon and headed for the kennels to retrieve Duke.

Duke was asleep when I reached his kennel.

"Wake up, boy. We're going for a ride on a big ole airplane," I said as I bent over to enter his kennel.

He was glad to see me, wagging his tail and circling me with excitement as I gathered up his gear and got him ready to go. “Easy, boy. Hey, glad to see you, too. Good boy, Duke. Good boy.”

Before loading Duke on the truck I took him into the exercise area, removed his muzzle and leash, and gave him the chance to relieve himself. He took off, nose to the ground, covering the whole enclosed area in a matter of minutes.

Sleeping, eating, and out of their cages for another all-nighter of walking and more walking. I'm realizing how little life these dogs get to live.

I walked Duke up the ramp and settled ourselves next to the cab to wait for whoever was going to take us to the terminal. The driver who showed up was Sergeant Wright who introduces himself simply as “Mike”.

“Moore, you are one lucky guy to be able to get off this base for a couple days. And going to Saigon. Boy, I'd trade places with you in a New York second,” he said. He climbed into the cab of the deuce-and-a-half. “I've heard the women there are really something else. You're going to check out the ladies aren't you?”

“I don't know…until you said something about it, well, it hadn't even crossed my mind.”

We got hit last night, Duke's here to be put down, and I'm thinking how hard it will be to get Junior out of his kennel and ready for work. Gals, not on my mind.

“I've not had very good success with the locals,” I said. I stroked Duke's ear.

“I know you're glad to be getting rid of that dud of a dog,” said Mike with a knowing grin.

“You're right about that, it was scary as hell last night. I found a hole and hid most of the night. I even started to get mad at the dog. It gets old real fast not having a dog to depend on, you know what I mean?”

“I read you loud and clear, buddy,” Mike said.

My name was on the roster and that was enough, no ticket needed. I found a seat in the small terminal area away from the rest of the passengers.

Doing something totally different from my normal night-to-night routine felt like freedom.

Just as the announcement was made to load up the little Caribou aircraft, a crewman walked over to me and motioned me to go ahead of the others in line and get onboard. Men moved out of the way to let us through and their show of respect gave me a lift I had never felt before. They had no idea that the muzzled Duke was a hippie dog who loved everybody and because of that he was on his way to Saigon to be put down. As far as they were concerned, Duke had bad hips, not an attitude problem.

We took off on time and quickly reached cruising altitude. I found myself looking out the window thinking about being shot down and taken prisoner. *Here I am again, in the middle of Vietnam during wartime without a weapon. Makes me wonder who is running this show.*

I was glad to see a few Marines and Soldiers with M-16s. An Airman sat next to me. He carried a sidearm, but he wasn't going to get far with a six-shot .38 pistol.

This airplane sound reminded me of a noisy insect/bug. The C-7 Caribou was a specialized cargo and takeoff plane. The aircraft was considered a workhorse and a lifesaver for the hard to get to and supply troops in the field.

The Airman sitting next to me saw I had no weapon. "Haven't you heard the stories of what VC do to unarmed Airmen? Being unarmed means you don't respect them."

I imagined a crash site where the same Airman next to me was going around and shooting all the unarmed Air Force personnel in the head.

"I'd rather you shoot me in the back of the head than to be captured," I said.

Duke rested between my legs just like he would do on the relief truck. The cool aluminum flooring must have felt mighty good to his hot paws. I gave water to the thirsty dog and he drank it all.

Once I told them the dog was being put down for medical reasons the conversation sort-of died out. But the word got around the plane like good gossip on fire. One of the pilots left his seat to approach me and Duke. He stuck out his hand to shake mine, and that was when Duke went for his balls and knocked him back into the flight cabin.

"Out, Duke! Out, Duke, out." Duke immediately returned to his sitting position between my legs. "Down, boy. Stay."

I had the biggest shit eating grin on my face but the pilot and half the men on board were bug eyed. I apologized, telling the officer he had caught me off guard. I told him he shouldn't approach a Sentry Dog without warning the handler first.

"Airman Moore, is it?" said the Lieutenant.

"Yes, Sir."

"I just wanted to thank you and all the handlers for the fine job you guys do guarding the officer's compound. Doesn't matter if it's raining buckets or there are sixty mile an hour winds, K-9 is there. We sleep easy because of you and your dogs, and I just wanted to let you know that you are very much appreciated. I mean that."

"Thank you, Sir. I will pass that along to the men. Believe me, we don't hear such praise often enough. Thank you kindly."

As we circled Tan Son Nhut Air Base, I got a chance to see Saigon from the air and to noticed how the Air Base sat right up close and tight next to the city. This had to be a security nightmare. I was thankful for our little puddle of an air base sitting in the middle of a free fire zone.

The Way It Is

The United States used Tan Son Nhut Air Base in southern Vietnam as a major base stationing Army, Air Force, Navy, and Marine units. During the 1968 Tet Offensive, there were a lot of Air Force Security Police casualties on this base.

It's different when you have civilians tripping over the fences, bunkers and minefields. Phu Cat had civilians, but nothing like it was here, a clusterfuck of a nightmare to protect and worse to defend.

A three-stripe Sergeant in crisp, starched utilities rode up in a dusty jeep to pick up Duke and me at the terminal for the ride to the veterinarian. Duke sat up in the back seat like he owned it. I tightened my grip on the leash and prepared myself just in case the dog got carried away in the excitement of the ride and wanted to jump and make a run for it.

This place was massive with nonstop activity and the noise was almost unbearable with choppers starting up with that familiar "whop, whop" as the blades cut through the already sticky, smelly air. I don't think I could ever get used to working in the heat of the day. My driver liked to talk and pointed out the sights as we drove along.

He got my attention when he showed me a blown-up bunker that had never been repaired. Four Security Policemen were killed there when a Vietcong sapper ran into the bunker with a bomb strapped to him.

"I remember being told in K-9 School about this bunker and here I am with it in my face. I wonder why the Air Force hasn't repaired the old bunker complex," I said.

My driver informed me, "The Air Force wanted to make a statement to all incoming security forces to always be on their guard," said my driver.

I was in awe of all the activity. There were commercial jets waiting for takeoff as well as long lines of transports heading for all destinations in Vietnam. I watched the colorful freedom birds wondering which were taking home a load of armed forces personnel to the world and to life.

I could see a C-141 Starlifter parked on a ramp and my heart almost stopped beating, they're even more huge up close. The Starlifter could carry 154 passengers, 123 paratroopers or 80 litters for wounded with seating for 16.

I wonder how many coffins this giant plane can carry?

The big bird with its open belly, ramp down, was waiting to be loaded with flag-covered coffins for the long journey back to the world. There were several hundred coffins sitting in the hot sun waiting to be loaded on the giant planes.

I wonder why the big display of our dead? Right in the middle of the hippie heyday, good ole Uncle Sam is showing the whole world its freedom-loving humanity, and our blood-soaked dead brothers are roasting in scalding hot coffins. Creepy shit for a ride home.

The next thing my driver pointed out to me was LBJ. "You mean Lyndon B. Johnson, the President?" I said.

"No, this LBJ stands for Long Binh Jail. In between the walls of this high security prison for American GIs, concertina wire has been crushed down with layer after layer of wire until the interior area between the walls is nothing but wire. There is no chance of escape."

"I never heard of LBJ before seeing this place, but I won't soon forget it."

Duke was taking the ride in stride and was enjoying the air across his face like any red-blooded American dog would. I'd be glad to get the muzzle off him so he could rest a little easier.

I felt like a tourist as my personal driver and unofficial guide showed me the sights. The base was huge and people were moving around like it was a winter day in New York City.

"See that building over there?" the Sergeant said. "That's the morgue."

A large two-story building with no windows was off to my right. This place looks like cold storage for cattle or furs. The Sergeant started to tell a strange tale about dead bodies being stuffed with drugs before their return trip to the world.

I looked at him as if he were mad. "You're bullshitting me?"

"No."

"Why are you talking shit like this?" I was mad. "You're telling me that those coffins we just passed, those bodies in those coffins are transporting drugs back to the world."

"That's right, Airman," said the driver. He had a crooked smile on his face. "The rumor has it that the CIA is running drugs from the Golden Triangle to the World right through this place. How's that shit grab you? Tell me that's not fucked up."

"What is being done about it, Mr. Security Policeman?"

He looked at me with a mean sneer on his face. "Not a damn thing, not one damn thing." He spat. "Just the way it is in fuckin' Vietnam. You know what I mean?"

"Hell, no, I don't know what you mean," I shouted.

"All right, all right buddy. Just chill, will ya? It's no big deal. It's fuckin' Vietnam so chill out."

By now I was so pissed that tears were coming to my eyes.

"Look man, it's just the way it is here. There is so much money being paid out to everybody that nobody gives a shit anymore. Sure, every once in a while, we will investigate and make a raid on this place or a shipment of coffins being loaded up for transport stateside, but nothing comes of it."

"You are not kidding me, are you?" There was a long pause while I waited for his answer, but he was still silent, so I added, "Man, your messing with my mind."

"We can't even get into the coffins once they are sealed, and that, my friend, is the end of the story. Big money and big people are involved. Little people like us get to pay with our lives and pay again on our ride home, in a hot-ass coffin, of course."

The thought of four dead security policemen and of their guts being stuffed with drugs to be sold back in the States was too much to handle. "Fuck this place. I can't believe this kind of shit goes on. Fuck Vietnam."

Duke will never see the inside of this morgue for humans or take a last ride in a Starlifter going home. Instead, he will be buried in Vietnam and who knows what that means? His body won't be stuffed full of drugs and that's something, isn't it? Duke doesn't deserve to die. What if they sell his body to some Vietnamese restaurant to spice up some fried rice and noodles? Oh shit, shut up Jimmie. This place is driving me nuts.

At last, we arrived at the veterinarian. The familiar sound of barking dogs was a relief for me, but poor ole Duke didn't have a clue to his fate. I cracked the door and announced a "muzzled Sentry Dog entering."

I got an all clear to enter and felt the cool air-conditioning on my face as I closed the door behind me. The small room was spartan with just a small desk and two chairs in the room. There was a phone on the desk and an Airman First Class sitting behind the desk. The AFC wore large, rimmed glasses, a big smile, close cut hair, and a bleeding pimple gave me the impression this guy was either a reject from dog school or from a high school ROTC program.

I stood silent, waiting for Mr. Pimples to look up.

"Hi, Airman. Welcome to Saigon. Are you Moore from Phu Cat and is that Duke?"

"Yes, it is," was all I could get out. I wasn't about to say "sir" to this pencil pusher.

"Fine, we have been expecting you and we have a place set up for Duke."

The Airman relieved me of Duke and I followed him through a door into a room big enough to hold kennels and several tables including an operating table. Right off I didn't like this place; every wall was sterile white and it smelled of antiseptic like any other hospital would. I did like the AC, the first I felt in a while. I was relieved of Duke and he disappeared without a single hug or goodbye behind some double doors. Damn, I could hear dogs barking and wondered at their fate.

The Veterinarian, Captain Martin, sat behind a desk covered with papers, a telephone receiver to his ear and a picture of his wife or girlfriend just behind his head on the back wall. He also wore dark rimmed glasses, but his hair was jet black and his eyes were green. He finished his conversation on the phone and turned his attention to me.

"Come in, Airman." He said.

"Sir, Airman Moore from Phu Cat Air Base reporting as ordered." Snapping a salute, I used my sharpest military decorum, hoping I still remembered how it all went.

"At ease, Airman, and take a seat."

Captain Martin instructed me to be back at 1000 hours the next morning and that I was free to go. The driver would take me to the Security Police Squadron and get me bedded down for the night and loaded up for the ride to the SP barracks. I was ready for some chow, a bed for the night, and possibly some time in Saigon proper. The thought of women did not enter my mind. I hoped hippie Duke was comfortable.

During my ride around this base I noticed how crowded the barracks were to each other and the rows upon rows of them as compared to Phu Cat. Mostly weeds grew here and when the wind blew I got a mouth full of red clay. There were no "Keep off the grass" signs here.

Saigon, Here I Come

I was in the big city, a city the size of New York, so I wanted to check it out and get some idea what it was like to be in urban Saigon, to see the people in action. I was hungry and a little luck arrived as an Airman named Theo asked if I would be interested in a good steak dinner in town. He was headed downtown and I was welcome to tag along.

The bus ride got interesting as soon as we went through the main gate into Saigon proper, where all of a sudden, there were millions of Vietnamese people coming at me. There were few cars or trucks except for military. Bicycles competed with mopeds and people on foot ran for their lives. You could see the exhaust from the traffic that hung in the air like fog, but no one seemed to pay it any attention. Some people wore masks over their faces to protect their lungs from the nasty air.

There were also plenty of Americans in and out of uniform. Military equipment in convoys was on the move everywhere. A few mud-caked M-48 Patton tanks were sitting stationary, with South Vietnamese ARVN standing guard.

The street colors were brilliant and pleasant to the eyes. I focused on every woman that came into view. The Vietnamese women of mixed blood, as in French and Vietnamese, were drop dead beautiful. Those with Chinese/Vietnamese heritage were taller and attractive, but the classy girls were the home grown, high society or college type Vietnamese women. They dressed in all colors of silk pants with a split one-piece dress. They were completely covered up with no skin showing except for their face, neck, hands and ankles.

As hot as it is, they cannot be comfortable, but they look cool and fresh as only a woman can do. These skirt things they wear, when they blow in the wind it looks so very sexy and for damn sure, stimulating. I would love to find a woman who would just screw me, out of my mind, take my money and invite me to come back.

I'd love a good hamburger and I was used to eating hamburger, not steak, so it was with the help of my new friend Theo that I ordered.

"You can't go wrong with the New York Strip," Theo said.

When the steak came. "It's burnt," I said.

"Don't worry, you'll love it."

Theo was right. I ate it and wanted another. We moved outside to a table on the sidewalk under some shade so we could do some girl watching.

I was not disappointed. I got an eye full and a whole new perspective of the people. The steak was delicious and I did order another before I left. I looked down the street, I was surprised to see a large Coca-Cola sign and what appeared to be a bottling plant. I ordered a Coke.

"This is what the natives drink. Prepare yourself to be disappointed," Theo said.

"It tastes god-awful." I spat. "It's as flat as a local's chest."

"I warned you it's not worth the fifty cents."

I didn't bother to finish the Coke. Instead, nothing but an icy cold beer on a hot day in Vietnam would do as my second steak arrived.

Along the street I noticed that every soldier was carrying a weapon. The only people without guns to shoot were Air Force people.

I feel naked.

I hadn't forgotten the bombing of our K-9 bar back at Qui Nhon. It's so easy to kill people if you really want to. I told Theo of Qui Nhon and how the town was put off limits after the VC blew up the K-9 bar. Theo said there were things being blown up in Saigon every day, but the whores of Saigon were second to none.

It had been a long night and a longer day with no sleep, and I was beyond tired. Theo bid me farewell, and after giving me instructions on getting to the bus stop, he left for some girlie action.

I had lost interest in paying for sex. Nothing about it felt right. After my experience with the women of Qui Nhon, sex became secondary to surviving. I would stick to jerking off, whenever/anytime I needed to, with a Playboy or Penthouse, it was safer.

When Theo left, I sat for an hour or so drinking at least a six pack of beer trying not to dwell on Duke. I was happy to sit, watch girls, and feel the vibration of this faraway Eastern city. It took a war to get me to experience a strange new culture. I liked it a lot. Then I heard a slow roar coming closer and an Army convoy came into view with lots of trucks and Patton tanks. It broke my stupor.

It's getting dark and time to go to the kennels, get my dog, and go to work. No, wait, this is Saigon.

Instead I found the bus stop and waited for the return trip to the Air Base and safety. As the blue Air Force bus made its way along the crowded streets, my thoughts returned to my hippie dog, Duke.

Why is it necessary for me to be at the Veterinarian's clinic in the morning? Maybe I need to fill out some paperwork. There is always paperwork in the Air Force. I hope I won't have to see Duke again.

I skipped evening chow and hung out around the barracks watching TV and talking with some of the off-duty dog handlers. Before I turned in for the night, I took a walk around the barracks area. Security on this base was different, and the tension felt by these airmen was obvious. Just like Phu Cat, flares dotted the skies and I couldn't help wondering what those flares meant.

I went to bed, and slept like a dog. Nothing would disturb my sleep. It was 102 degrees that night and I slept through the heat as well as the rain that came some time later. But by morning, it was sunny and already hot as hell. I ate a big breakfast not knowing when I would eat again. The same Sergeant was waiting to take me to the Vet and then on to the terminal for the flight back to Phu Cat.

Friend Fairwell

The next morning, I reported as ordered and prepared myself for whatever came next. This time a Sergeant greeted and escorted me to an area of the hospital where I had dropped off Duke. The vet was seated at a desk and upon seeing me got up and led me to a corner of the building where Duke was lying on a blanket. He lay on his side and looked like he was asleep. As I sat down next to him, he lifted his head and look at me with far away eyes. He greeted me with a couple wags of his tail, then dropped his head down and remained still.

The hair around Duke's neck had been shaved off. A tube with a needle was taped to his neck. I felt sick.

My inability to get this dog to work is going to cost Duke his life. For better or worse I am the reason that this dog is going to die. Shit! I'm about to lose it. Stop it, Jimmie! Stop crying. How embarrassing.

I spoke out loud. "This is bullshit. The dog is a dud and he could have cost me my life. Fuck this shit. I've done nothing wrong. Satan get thee behind me, you asshole."

Where did that come from?

I felt the need to explain myself. "I feel so guilty. He's such a good, strong dog and I had hopes that he could be returned to civilian life to be someone's pet."

The Captain looked at me and I could see in his eyes that he understood what was going through my mind. "Sentry Dogs cannot be reintroduced back into civilian life. They are too dangerous. None of the dogs serving in Vietnam will return stateside due to viruses that infect most of the dogs."

The horrified look on my face must have said it all.

"Airman Moore, remember where you are. This is an insane war and there is no time for sympathy for an unproductive dog. When something doesn't perform to standards, it is replaced. We replace broken dogs the same way we replace a broken radio or a broken M-16."

"I understand what you're saying, but I've worked him almost eight weeks and have become attached. Even though I was scared to death every night I stepped off onto my post, still I am very fond of him."

"Yes, it's a waste of dog flesh," said the vet. "Airman, look on the good side of it. You are here because a dog became available, and from what we're hearing, a very good, brave working dog."

"Yes, Captain. Thank you."

"Forget it, Airman, it's nobody fault."

I was feeling better about the whole thing until the vet asked if I would hold Duke and pet him as he administered the serum.

Hold Duke in my arms as you kill him? Are you shittin' me? Fuck me.

I went to my knees, lifted Duke's head into my arms, *imlosingit* and held him like he was a newborn baby. He opened his eyes. I could see his tail move several times as he stared with a faraway, glassy-eyed look. My heart beat faster and faster as I held him.

Duke was hooked to a machine by a thin tube. I watched as the Vet released the pink fluid that moved from the machine toward Duke's neck. Duke was going to die in my arms.

I couldn't breathe. This senseless murder is wrong. "hudamightynos", Pull the needle from his neck, grab him, and run for his life. Stupid, stupid, stupid.

I watched as more and more of the pink death entered Duke's body and it stunned me to feel the life force drain from his body. I continued to look into Duke's light brown eyes until they faded away. The body went limp, the eyes closed, and I felt his body give up his spirit so that it seemed to go right through me. I covered my face with my dog's head so my tears wouldn't be noticed by the Captain who was still by my side.

He noticed anyway. The Vet put a stethoscope to Duke's chest. "Duke is dead."

That's when I lost it.

After a short pause he continued. "Thank you for coming back this morning and helping out. It's tough on you right now, but you, me, the Air Force, did the right thing."

"Yes, Sir."

I got up and walked outside, thinking the heat would offer some relief. I found a shady place to hide and finished my grieving for Duke. I had to go back into the Vet's office and face the men in there. I felt and looked like a cry baby.

Upon entering the building, I could see Duke lying in the corner. He would not jump up and greet me anymore. The Vet gave no quarter to my feelings.

Shouldn't he be covered, at least?

"Listen up, Airman; you're not the first handler to have to put his dog down. Sure, you were close to this dog and it hurts to have to lose him, especially this way. But from the written reports and information about Duke, you were just one of three other handlers who have tried to work with the dog."

"Moore, you are going to be all right. You will get over this. It is what it is. That's some dog you are taking control over, so get your butt back to Phu Cat and be glad you finally have a Military Working Dog who will keep you alive long enough for a return trip on a freedom bird."

He put out his hand and we shook.

"I'll be looking for you in a week or so," he said, "when I come to Phu Cat for a check-up on all the dogs. Good luck."

The Captain laid it on hard and it made a lot of sense. My ride to the airport passed the morgue and the stockade called LBJ and the bunkers where my fellow airmen gave it all. I was ready to get out of there. *Friend fairwell.*

Duke, duke, duke, Duke of Earl, Duke Duke; Duke of Earl, Duke, Duke; DukofEarl! And just like that Duke II exited my life, *pretty much for good.*

My Walk on the Wild Side

I stood in a long line to do what people do in airports all over the world, wait. It took over an hour, and when my turn came the news wasn't good. The next flight to Phu Cat Air Base was at 1830 hours. It was lunch time, so I would have to wait another six hours to catch this flight.

I asked the Vietnamese clerk about other options. "Good news," she said, "you can catch the flight to Qui Nhon leaving in about thirty minutes."

I figured this would be a smart shortcut that would get me back to my base several hours earlier. From Qui Nhon, I could catch a flight to Phu Cat. That shouldn't be a problem, and if not there was always the bus. The only flaw I could see in my plan was if I got back to base too early, I would have to go to work that night.

For the trip home, I wore my camos and stood out from most everybody on the plane. I looked around for weapons onboard and there were very few. With my dog gone, I had no weapon either.

I feel useless, dressed up and no weapon to play with. This is not cool.

It was a different flight on the return trip to Phu Cat and I was in no hurry to get back to my base. On the flight north, I felt the emptiness of loss; it felt cold inside, even in the heat. It had been just a few hours ago that I held Duke in my arms as his last breath faded from his body. I missed him sitting down on the floor between my legs. I missed being able to drop my hand and feel his hot breath against my sweaty palm.

He was gone and I am still left with ten months and eight days of duty to finish. What will that be like, working a real Sentry Dog and a hero? WOW!

Sleep in the rough riding Caribou was kind to me, and I didn't awaken until the small plane came to a stop. As I departed down the short ramp and into the stifling heat, I thought I was at Phu Cat AB, but I still had thirty miles to go. I went to the flight desk, only to find out there were no flights leaving for Phu Cat until early morning.

I could have taken some form of ground transportation but figured hitch-hiking would be an interesting way to see some of the countryside. I carried a gym bag with a few items in it, nothing else. I figured since the bus from the base made the trip several times a day without any problems, then I wouldn't have anything to worry about either.

Off I went, walking out of the small terminal building and then the main gate.

Highway One is right in front of me, but which way to go? Seeing the ocean in my face tells me that Phu Cat is the other way. I'll travel north for a little while and then west.

After walking for an hour, I stopped to buy a soda from a roadside vendor and took a break. I was still walking parallel to the security fences of the Army base. They used wooden crates stacked several stories high as a part of perimeter wall. There was no sidewalk between the highway and the base perimeter. The wall stopped and the road began.

Several times, different jeeps or trucks offered me a ride, but I declined because I wasn't tired of walking and wanted to check out the city. I received some strange looks from the Americans who caught my eye.

After several hours of hiking without water, I began to ponder my situation. I tried to enjoy the view of the countryside, where the only ugly in Vietnam was the scarred, burned out land, an after effect of bombings and God knows what. I came upon a small, tranquil village and the green of the mountains, hills, and valleys full of rice paddies worked by families of farmers.

It was all very peaceful as if war had not been declared here. I walked over a bridge and looked down at the brown waters of a small creek. The village was surrounded by low green mountains and the large expanse of rice paddies for as far as the eye could see. This was a picture worthy of National Geographic.

I lifted my hand in greetings to the few men I encountered, but there were no return greetings, only angry looks. They would stop and stare, eyes like angry tigers with their balls caught in a trap.

I had fucked up. If looks could kill, then I need to get out of here and fast. It may not be in my interest to continue my sightseeing.

I put my thumb out and caught the very next ride with friendlies, a couple of ROK troops returning to Phu Cat. As I climbed aboard the big deuce-and-a-half and settled down near the cab of the truck, one of the Koreans asked me why I was out here by myself.

"I just decided to take a walk on the wild side," I said.

They laughed at my response and looked at me like I was crazy. I shook my head and shrugged my shoulders.

They're thinking this is one crazy-stupid GI. I won't be telling anybody about my experience into stupid-land or the possibility of dying at the hands of some farmer by day and Vietcong by night.

The thirty-minute ride was enough time to make my stomach burn. All my boldness slipped out of me as I contemplated what my situation might have been if I had been captured, and what an opportunity for the VC to gain intelligence from a captured Air Force dog handler. I would never leave the base again.

I headed for the chow hall first thing. Then I caught the bus out to the kennels to see if I had to pull duty and check on Junior.

I walked into the kennel office to check the duty roster and to my delight I wasn't on it. Next I checked in with the First Shirt to let him know I was back. He wasn't at his desk, but the Airman in Charge said he would be back shortly.

"How was Saigon, did you get laid?" asked the Airman.

War and sex.

"Truthfully, I'm sick to death of whoring around and I never want to be with one again," I said, "but what I can't say about the Saigon pussy, is whether it has hair on it. Not that I got laid, because

I didn't, the nice fellow who took me into town for a steak dinner informed me."

"Shit, man, I'm getting a hard on just thinking about it."

"Look, you got me checked in?'

"You're cool, and you got the night off, you lucky fuck. You're on early shift tomorrow night."

"Guess it's time to see if Junior is going be man's best friend or make me a new asshole," I said.

The Airman in Charge laughed. "I hear ya, watch your ass."

Queer Introductions

The kennel area was like a ghost town, not a sound to be heard nor any movement to be seen. One or two of the dogs would open an eye or lift a head in recognition, to let me know I had been duly noted and to proceed with caution. One of the dogs was chasing something in his sleep. His legs spasmed with crazy movements, and he would make soft, whimpering sounds, but then the chase was over.

This morning I held Duke in my arms and felt his last breath leave his body. Again. The Grim Reaper was after my ass on Highway One but the surroundings had me smiling. I came close to dying in some shithole village or worse. What the hell was I thinking? I could've been headed to the Hotel Hilton in Hanoi and a bamboo cage not big enough to hold a dog, much less a man. Damn! It's been a long, interesting two days I'll not soon forget. I'm beat to the bones.

I should have married Lillian and stolen her away with me. Maybe she would still be living. Maybe if I had been a better handler, Duke II would be alive. I should have been there for both of them. I know, I know, God's will be done. "This too shall pass."

I already had my leash, choke chain and muzzle. I didn't have a collar and would have to go by Duke's old kennel and get it. I returned, retrieved my gear, and headed for Junior's kennel. I would ask permission to move Junior to Duke's old home; his kennel faced out towards the exercise area and the mountains. It was a nice view.

I noticed several of the kennel staff taking a smoke break under the shade of the picnic area.

"Hey, Moore, you going to take Junior out for a walk?" Russell Puls snickered. The surfer was on temporary assignment to the day shift, and it was cramping his style. He was used to spending his days napping by the pool or at the beach. "It's only been two nights – think the mutt will let you in the kennel?"

"What choice do I have? I have to have him ready to go tomorrow night."

What a life it must be to work the day shift. Being on post at night is safer than being asleep in a barracks when all hell starts to rain down. I count working a dog in the dark and in the rain as a blessing. We all do.

Junior was on his paws and staring at me as I approached the kennel door to open it. Junior slammed into the door with such a force that I almost let the door come open.

Oh shit, not a good sign.

I sat down and leaned against his door with my back to him and began to speak in soothing tones.

"Sergeant's gotta watch this. Moore is trying to get Junior out of his kennel. This may be very interesting," said Russell.

These guys know Junior and what he is capable of doing. Maybe they'll tell me he's a badass, or maybe they'll tell me Junior is a nice dog and probably won't bite me. Either way, it will be a lie.

Out of the corner of my eye I could see a crowd forming and couldn't miss Staff Sergeant Francis as they stood staring.

I'm not going to wimp out and let this dog make a fool out of me.

"Junior, my friend, you don't know how lucky you are to have me and me to have you. I've been waiting for you for six long weeks. I promise to love you, Junior, I do. In turn, I want you to keep me alive."

Junior was sitting now, sounding a low growl in the back of his throat.

"I will take care of you, Junior, I will. You will never suffer from lack of water, no sir, no way will I let it happen. I will bring you treats. I will brush out your coat every night, towel off your wet coat at the end of another long night in the rain. I'll even let you under my poncho and out of the rain. I promise, boy."

"Anybody want to make a wager on Moore getting that dog out of his kennel today, ha-ha, and walking him to the exercise area?" said Russell. "Hey, I'll put ten dollars down myself. Any takers?"

"Come on, boy, put your teeth back in your mouth and let me in." I ignored the hecklers. "It is time to get you out of here and to the exercise area, Junior. We have to work, that's working together tomorrow night without you attacking me."

"No takers, Sergeant?" Puls wouldn't let it go.

Sergeant Francis snapped at Russell. "Cut the crap now, Puls. Just watch and keep your mouth shut. We all might learn something."

I can't believe Sergeant Francis is coming to my aid.

I decided it was time to go in that cage, put the muzzle on Junior, and walk him out past the small crowd who were wanting me to fail. Ego is a powerful thing.

I will show them.

Junior was having nothing to do with my plan and he certainly didn't care about my pride. He was still in mourning for his lost handler and made it clear he wanted me to get lost.

"Okay, Junior, I can see you are not in any real hurry. In the morning, we will finish this, okay, boy?"

I put my leash, muzzle, and collar just inside the kennel door in hopes he might get the meaning that he had a new handler. I felt like I was losing face or something as it was time to head for the bus stop. First, I had to walk by Puls and the other guys.

"Moore, it's no problem, man," said Sergeant Francis. "You've broken the ice, and tomorrow you can get him to the exercise area. It's no big deal."

"Thanks, Sergeant."

I can't believe Staff Sergeant Francis is saying this to me. He's saving my face.

I considered that I had worked 45 straight nights, and this would be my first night off ever since arriving at Phu Cat. I went by to check out the movies that were shown outdoors if the weather permitted, but they were having trouble with the projector, as usual. When the show started several hours later, I had already left.

I walked by the new pool next to the new NCO club and entered the Airmen's Club. The place looked like a larger version of an empty barracks we lived in. A stage was at one end, there was a bar made of planks, about twenty tables on a concrete floor, and that was it. I had forgotten our chow hall doubled as our club.

There were a few men drinking and listening to music. I stayed long enough to drink a beer and left. The tired-ass lifer behind the bar never said a word, so I gave up on the night life at Phu Cat Air Base.

I stepped out into the night air and the first raindrop hit me in the eye when I looked up at a lot of nasty clouds. I started to run, and just made it before the bottom fell out. Our barracks was empty and not a sound to be heard.

I cut through another room on the way to my room, but out of the corner of my eyes I saw something strange in the darkness. I stopped cold in my tracks and took a couple steps backwards. My eyes adjusted, focusing on two guys in one of the bunks. I stood staring in disbelief, trying to catch my breath and get my thoughts together.

What the hell am I looking at? No one should be in here, everyone is on post.

"What are you doing in the K-9 barracks?" I saw the two men were undressed in one of the beds.

Whose bed? I don't know, but these two shitheads are in the wrong bed at the wrong time.

"What have I got here, a couple of homos?" I asked. "You bastards, this is a Security Police barracks and I'm a Security Policeman." The two stayed quiet, but they got up and started putting clothes on.

"Queers in the K-9 barracks. No way!" I did not recognize these assholes. "Stop, let's see your IDs now."

The Security Police office for the day shift people is right below and is always manned. I can get help if I need it.

"You two ever hear of LBJ?" I asked using an official tone. I was planning on having some fun with them.

"No, sir." They both handed over their IDs.

I pretended to be pissed. "Who the hell gave you permission to be in here? You guys are definitely not K-9. Why are you even in this building?"

I shouted, "Stop. Did I tell you to put your clothes on? Leave your pants down, and if you give me any shit I'll leave you naked while Security Police are putting the cuffs on."

They froze in place.

"You two shit bags are in a world of hurt. LBJ stands for Long Binh Jail. Does that ring a bell? It's the ugliest jailhouse on planet earth. Do you think they will take kindly to a couple of homos? I don't think so!"

There were two naked men standing before me, scared to death, and as if on cue both started to cry. "Sir, what are you going to do to us?" said the younger one.

"Haven't you been listening? You're going to be packed off to jail, hopefully never to be seen or heard from again. I can't believe you two shit bags are cumming all over a dog handler's bunk while he is out on the line protecting your sorry asses."

"I'm sorry, sir, it won't happen again. Please, let us go." The other guy just shook his head because he was crying so hard.

"Sorry won't cut it. I'm a Security Policeman and you've broken very serious laws. Homo fag behavior is unacceptable in the Air Force and a Court Martial will have you two in Fort Leavenworth for 25 years at hard labor, if they don't shoot you. You have royally fucked up."

They composed themselves. The one doing the talking asked, "Is there any way we can get out of this? Please, anything? I don't want to go to JBL."

"It's LBJ, scumbag. What are you offering, a blow job?" I was having so much fun with these two that it was hard to keep from laughing.

The talking one pulled out his wallet and offered it to me.

"So now you're trying to bribe me?" Then I said to the other one, "Well, speak up, how about you?"

He pulled out his wallet, opened it up, and the bills in there got my attention.

"So, you're suggesting I just hand over your IDs and let you put on your pants and walk out the door?"

"Yes, sir, please, you'll never see us again. We promise. Please sir, don't send us to jail."

"How much do you have? Count it up," I said. I waited to find out what money they had between them. I couldn't believe my eyes when they handed me over six hundred in script.

"Going to the worst jail on the planet is worth a measly six hundred. Are you kidding me? Don't you have any US dollars?"

"No, sir, we don't. Please take this money and let us go."

I should be ashamed of myself, but fuck these idiots. They got a lot of nerve coming into my home away from home to fuck.

"Don't you know Queerdom is a sin?" I said. "Don't you know what the good Lord says in the *Bible* about homosexuality being a sin and punishable by being stoned to death. Are you sure you've no more money? Better take another look."

"No sir, we are telling you the truth. Please take the money, it's all we have."

"Put your clothes on," I said. "You two get the hell out of here, and if I ever see you anywhere near this barracks I will have you both in chains. I see you now clear as day, so don't fuck with me."

I didn't want their money, so as they turned to leave, I gave it back. "You're Goddamn lucky I'm in a good mood today, I don't want to ever see you guys in here again, got it?"

I ran down the stairs looking for one of the K-9 people who worked the day shift. Of course, none of the lifers could be found. Next, I went to the Security Police desk which was located on the first floor directly underneath from where our room was on the top floor. Nobody was there either. I didn't know what else to do. It was raining cats and dogs and, by this time, I was tired, so all I wanted to do was sleep.

I took a long, hot, but nervous shower. The latrine was quiet except for the rain and me. I had to admit being a little spooked from my recent encounter with those queer airmen. As the heat of the water relaxed my body, my mind was trying to come to grips with questions.

It looks like anybody could use this place as a whorehouse. Sergeant Francis' naked body getting it on with mama-san popped into my mind. *Gross!*

When I dried off it was now deathly quiet, the rain having stopped. I stood listening to the night, glad the lights were on as I headed for my bunk. I put my K-Bar knife under my pillow.

The lifers would have written them up in a heartbeat, and they would have ended up in jail for a very long time. Those two homosexuals will find somewhere else to play and thank me for letting them off so easy. Their money ain't shit compared to 25 years in Leavenworth, Kansas.

Sometime in the night I awoke, hearing something leaving the tube at the mortar pit and heading down range, probably to light up some poor, soaking wet handler and his dog. I could hear them in my mind, cussing.

What is going on out on the perimeter? I hope it's a flare and not a mortar.

Five minutes later, another round was sent down range. I laughed to myself as I thought about all the handlers and their dogs having a long night, already soaked to the bones. Then the night became quiet as the slow pitter-patter of the rain began again. The rain came harder and I slept hard.

A million flashes leapt through my mind from the last couple days of fast living. In my dream, flares lit the night sky while I held Duke in my arms and felt his spirit leave his body. Junior refused me entry into his kennel. I walked again through the pristine village seeing death in the eyes of farmers.

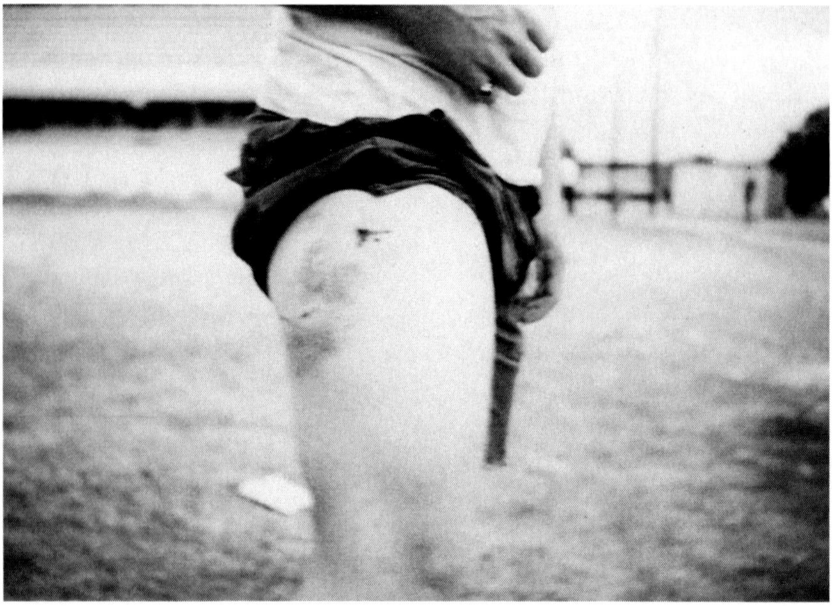

Chapter Eight

JUNIOR

Phu Cat Air Base, Vietnam – Late February - April 1969

The first thing on my mind upon waking was not breakfast, but facing Junior in his kennel and putting a muzzle on him. I was nervous for sure, not scared. Junior had so much of my attention that it took a minute to realize a game of cards and drinking were happening three feet from me. I must have been sleeping hard not to have been awakened as these guys came in.

"Jim."

"Who's asking?"

It was Vinny. "Jimmie Moore, wake your sorry ass up."

"W-wake up, M-moore, get your r-rear out of bed," said Joe.

Barry chimed in. "Jim, get up and come play come cards."

Joe, Barry, Vinny, and Mike Schmitt were all hollering at me.

"It rained all night and I thought about you lying up in bed while I'm out there getting soaked to the bone," said Joe. "I almost started to hate you."

"All of you can fuck yourselves. I enjoyed listening to the rain all night long. It was dry in my warm, comfortable bed, and the pillow was nice, too."

No one heard my ribbing because lucky Joe won yet another hand of poker. As the screams and obscenities filled the air, all the players were up on their feet giving Joe all the crap he could handle. Joe lit up a cigarette and just smiled.

Bolden shouted from his bunk that he was trying to sleep. This was seconded by a host of others also trying to sleep.

"Joe, hey Joe! You want to eat some breakfast?" I said. I put my shoes on. "I'm headed there now, so get it together."

"I pass, man, I'm tired. I had a rough night with d-damn trip flares getting set off by the rain, I guess. Who the hell knows? Damn, Charlie could be p-playing games with me. I had two I had to check out by walking up to them in the light to see whether Sir Charles had come knocking. I'm telling you, man, this sh-shit gets in my head. I'm a nervous wreck. Need to sleep it off."

Then Joe yelled out to all. "Winning all your money this morning was a great help and I thank you, I'm feeling much better."

That being said, the place lit up with cursing, Joe this and Joe that.

"Jimmie, I'll go with you. Just let me get my shoes," said Vinny.

By the time we left, I was accompanied by Vinny, Teagan, and Mike. They talked up a storm about the rain that lasted most of the night, how miserable it was trying to stay dry, walking in the mud and falling down, floods of water heading down hill directly at them. For now, the sun was bright and hot, and the day had already begun to stick to me. Pretty much same-same.

When it came time to visit Junior, the morning was still humid at near ninety-plus, but it seemed cooler or less warm or maybe I was just getting used to the heat. The air through the window of the blue bus was cool on my face like air-conditioning.

I didn't announce myself to anybody but walked past the office and entered the kennel area. The dogs were on my side, not one was up or barking. It really had been a rough night, so the dogs were exhausted and sleeping.

I opened the gate to Junior's cage, grabbed the leash and muzzle, but Junior didn't move or bark at me. He just lay there looking at me. I stopped and waited a few seconds.

"Good morning, Junior. You're a good boy, Junior. Boy, it's now or never, you going to let me in or what?"

I held my leash and the muzzle I used with Duke up to Junior and he seemed calm. He wasn't barking, just lying there watching me. I waved the muzzle and leash and again, no interest. When he looked away as if bored, I made the decision to take a chance.

I was locked in with Junior in his kennel. He just looked at me as I closed the gate behind me. That's when the barking started and the teeth came out. I sat down on one knee, reached around and unlocked the gate for a quick escape if needed. Junior just barked and growled in a bloodcurdling style, until he stopped and looked at me, snarling and teeth bared. I shut the gate again, readjusted my balls in my pants, and got to my feet. Junior never took an eye off me, but he had stopped barking and put his teeth back inside his head.

By now, at least fifty dogs were up and barking. This didn't help my cause at all.

I locked the gate behind me, never taking my eyes off Junior for a minute. He took a couple steps towards me and started barking again. I continued talking to him and sat down on the cement floor, at his mercy now.

While Junior was in the sit position, I took a long minute to examine my new dog up close. His coat was yellowish red and thick but not too long. His fur was thicker and longer than Duke II's, and I knew this called for smarts on my part to keep him hydrated.

I tossed my leash at Junior, causing him to back up. He had stopped barking and began sniffing at the leash. I noticed his big, pointed ears and light brown eyes; he was a smart looking dog. Junior was a compact dog for a German Shepherd, meaning that he was neither a long body and close to the ground as most German Shepherds are, nor was he tall and lanky like Duke II was.

If only this dog will cut me some slack and let me put a muzzle on him.

Next, I threw the muzzle at him and it bounced off his snout. As I got up and grabbed ahold of the leash, for a spilt second of time, fear came on me and that dry, sour taste in my mouth was offensive.

I'm so thirsty, but I have to go for it. No turning back.

I saw Junior was up but backing off, not barking, just waiting on me to make my move. "Hey, new handler, you got the balls to take me on?" I imagined Junior to say.

I retrieved the muzzle, then moved up to Junior's side and slipped it over his black snout. "You're a good boy, Junior, yes you are." Then I made the muzzle snug, not too tight, behind his ears and buckled it. Next, I attached the leash to the choke chain and we were ready to go. He seemed to respond to both praise and loving up.

I decided before taking him to the exercise area, I would show Junior his new home where Duke had been kenneled. Once inside with the door closed, I took a few minutes to brush him down. I was looking for the tail to start wagging and it did. Junior began to enjoy the attention very much, and I knew I had him.

We prepared to leave the kennel area and that's when it happened – he jumped me. I was so surprised that I stumbled backward, nearly losing my balance. The adrenaline rushing through my body was fierce as I hit the chain link fence which kept me from falling on my butt. I about shit in my pants, but to my great relief, Junior was licking me through his muzzle.

Thank God. False alarm. "Ah, Junior, I love you, too."

He got down on all fours and started to scratch, which was a good sign that the introductions were over and Junior was ready to go enjoy a little freedom in the exercise area.

"That's a good boy, Junior." With any new dog, a handler talked and kept talking until they both came to an understanding of who was the boss.

I know I still have to return him to his kennel and get out safely, but first we will spend time running around the exercise area and see if he is willing to take commands from me. Next, the muzzle has to come off while running around and giving him commands. I'll know all I need to know about Junior in the next fifteen minutes.

Junior and I walked out of the kennels and into a bright sun, all the time I was praising him and talking to him.

"Junior, you're one fine dog. I'm one lucky handler to have a fine dog like you, a real hero."

I think the secret is to say his name a lot, praise him up continually, taking away his own nervousness, and hopefully, some of that aggressiveness.

"Let's go, boy." Clapped my leg. "Good boy, Junior." Moved out.

Now I was petting him and praising him like some crazy man. I felt good that this animal was accepting me and not wanting to bite me. As we marched, making left turns and right turns, Junior fell right in line. When I stopped I didn't have to tell him to sit. It was automatic and he even squared up to a right angle.

"Damn good dog, Junior."

A small audience had gathered at the kennel office area to watch us, like spectators at a football game waiting for the quarterback to be hit hard and not get up. They watched for mayhem on the obstacle course at my expense and a good story to tell at tonight's guard mount.

I paid them no attention at all. I walked down the grassy slope past the grill area and entered the gate to the exercise area, secured it and walked to the furthest corner and stood at attention to buy some time. I intended to make my thrill seekers stand out in the hot sun for as long as I could milk it.

I put Junior through a couple of right and left turns and about faces, heels, sits, until I was getting bored and I knew Junior would be heating up. I still needed to run him around the exercise area just once before taking his muzzle off.

Being thirsty and a little winded might take Junior's mind off me, so I'll take a chance and remove his muzzle. Will it work?

We came around the corner and I halted, Junior doing likewise came to a sitting position at my left side.

So far, so good. "You're a good boy, Junior."

I petted him as I reached under his neck and unbuckled the leather, holding the muzzle securely around his neck and behind his ears. The muzzle slipped off while I continued to praise him. Only two seconds passed before we were off and running. I kept a quick pace, which, of course, didn't faze a dog like Junior at all. We completed two running turns around the exercise area.

At the start of the third pass, I slowed almost to a walk and prepared Junior for running the obstacle course. The first was a ramp that he would run up on, cross over on a narrow walkway, then down the other side. Next was a ten-foot tunnel, and Junior didn't slow down as he crawled his way to the other end. I had to run faster to reach the other end before he exited. I was waiting for him and grabbed the leash as he ran out the hole. Next was the barrel climb that stood three barrels high, like a pyramid just over six feet tall. Junior attacked the barrels and was over them in a flash. A few more jumps and we had completed the course.

My fans had departed and I was in my happy zone. I gathered my gear and headed for the kennel, not putting a muzzle on the dog.

Oh shit, I can't believe I forgot! Is Junior really a man-eater that needs a muzzle on him? If no one is around the kennel, there is no need for a muzzle.

I talked to Junior as I walked him into his kennel, removed the collar, replaced it with his choke chain, and put my gear away. I had my back to the dog all this time and never even thought about it as I reached over him and got the water bucket with the intention of getting fresh water for him. I left the kennel, refreshed the water, and returned, putting it down where the sun wouldn't warm it up.

"Good boy, Junior. Thanks for not making a fool of me today."

I left to catch the bus, but first, NCOIC Master Sergeant Johnson called me into his office for a little talk. He asked about Saigon. "What were the women like?"

"Beautiful," I said.

"What was the base like?"

"We are very lucky to be at the Cat," I said. "It must be difficult to defend the base with it being backed up to the City of Saigon. We have it made here."

The NCOIC wanted the details about Duke and how I felt having to put a dog down. I really didn't want to talk about that and spoke as little as I could get away with, putting up a front as if I didn't really care one way or the other.

"Airman Moore," he said. "No other handler had any better luck with him, either, so don't let that get you down. Nice job today. You should have no problems patrolling your post tonight."

Before leaving I checked the duty roster for my assignment tonight. I was on the early shift, posted on Kilo 21, just another barren post with little to conceal a snake, much less me and Junior. It made no difference. I was excited about Junior and going back to work with a real military working dog.

I waited for the bus, feeling ten feet tall at the compliments from my First Sergeant and having the best damn Sentry Dog in Vietnam. I was ready to get back into the swing of things.

By the time I returned to the barracks it was almost thirteen hundred hours and the place was quiet. There were the soft sounds of music playing, so I knew brother Bolden was still up and listening to soul music on his small record player.

"Hey Mark, what are you doing up?"

"Jimmie...! You take care of business? Get your new puppy out of his cage?

"Yeah, man, wasn't a problem once I quit my fear, just walked right on in, put the muzzle on him and off we went. I had a little gathering at the office waiting to see me get bitten, but it never happened. Worked him about forty-five minutes, put him back in the cage with no problem and here I am."

"Good for you, Moore, you've got one of the best dogs in the litter."

"I know, and I'm excited being paired with a real working dog."

"How'd it go in the big city?"

"No fun having to watch your dog dying while you hold him in your arms."

"You had to hold him? What is that all about?"

"I'll tell you later, I need to get some sleep."

"Hey, Mark, could you play "Little Green Apples" for me? It puts me to sleep." It was a favorite to all of us because it meant so much to Bolden.

"No problem."

Getting high helps any music sound better. Here in the Nam, we are on an adrenaline high most of the time. We start coming down when we hear the relief truck coming, but back here at the barracks, there must have been a heck of a card game this morning because there's a stack of cans on the table almost to the roof. That's a lot of beer. Guys also drink rum, scotch, bourbon, and vodka at eight o'clock in the morning. You know, stress relief.

I finished off my beer and would have welcomed sleep, but I was too hot, and my mind wandered to Lillian, then to other faces of some of my school mates, until I had a clear visual of Jimmy. He and I were friends during our time as Scouts in Troop 70. We also had a ball playing drums in our high school marching band for years.

Even though a gentle spirit, he joined the Marines. Like a lot of the Kingsley Lake boys who wanted to go to Vietnam, said he was not about to be drafted or left behind by his friends. He was killed not long after arriving in Vietnam in 1968.

I was working at Jimmy Moore's Sun Sports (not related) the day that Jimmy's brother Tommy, with his mother and father came into the boat shop to purchase a boat for their one remaining son. The family had decided to use the military insurance money they got for their son dying in Vietnam to get a new boat for him. I wondered what local gossip would make of that. Tommy was a good friend; we drummed together in high school band. I wanted to go to him, he looked so sad but I couldn't move. He had already picked out the boat and was here to pick it up.

A horrible thought occurred to me as I recalled that day. *A man's life was only worth a new boat or ten thousand dollars. How cheap the government's bounty is on a serviceman's life. A boat for a life, fuck Vietnam.*

Then I began to get pissed off, knowing how ungrateful the citizens of America were and didn't give a shit. I started to hate all the messed up news coming from home. The hippies, the riots, the never ending bad news displayed for the world to see, and always the stupid government making life hard. Just like other handlers, I went deeper inside to get away from this God-awful mess I'd gotten myself into. Drinking was the only real pleasure or release we had.

I could hear Vinny and Joe in a heated conversation. Trying to sleep was useless. Eavesdropping while putting on my boots, I heard Vinny describing the girl in question as a slut.

"Fuck you, s-scum bag," Joe hissed at Vinny.

"You guys headed for some chow? I'm mighty hungry. I've got to get some treats for Junior. He's gonna' love pork chops."

"Oh, shit," Vinny said. "You're a different handler these days. It amazes me how you got Junior up and running in, really, just a day, right? It's been three days since we were hit, not bad.

"Damn straight, Jim," Joe said.

"I worried about it at first," I said. "I took him out for exercise for about an hour and had no problems with him. I've a good working dog and on top of that Junior is a proven war mutt. I might let him bite your ugly asses."

A New Beginning

"At ease, men," said the short, beer bellied Staff Sergeant. "Things have quieted down enough to lower our alert status but that doesn't mean you can relax. Stay alert. Also, Airman Lane was flown to Japan this morning and looks like he will keep his arm. We will keep you posted. But it looks like he is done in this man's war and headed for the world."

The Sergeant wheezed and launched into part two of his status report. "Good for him. His dog Junior was issued to Airman Moore, who will be taking Junior out for the first time tonight. Let's wish him good luck and hope he doesn't come back in the morning with bites all over him. Nah, just kidding Moore, but watch your back and good luck."

There was quick laughter and a few shot glances my way, then back to business as the Sergeant carried on.

"In appreciation for all your hard work and long hours the brass has okayed for us to have a party. They are paying the bill. Next Tuesday morning right after you put your puppies away, the grill will be going, the beer will be at near freezing, and the party will be on. There will be a little treat for any and all FNGs. What's it called, men?"

"Initiation!" the formation yelled. A cheer went up and a lot of clapping hands clattered. "Hot damn, a party." Everyone was whooping and hollering and carrying on.

"Alright Airmen, at ease…AT EASE, thank you."

I wonder what a party and initiation will be like in the Air Force. I've been told since day one that you are not a dog handler until you get bitten either by your own dog or someone else's. Are they going to turn a dog loose on me? Nah, they can't do that and get away with it. I wonder what happens if one of us chooses not to show up for the party.

I tuned back in to Sergeant Francis.

"Nobody in the Air Force died last night, let's be grateful for that. There were several planes destroyed at Bien Hoa. It was a mortar attack with no enemy KIA. Things have quieted down all over the country. Be alert, men, and live. We all want to go home after this bullshit."

"Junior," I said. "Get up boy, it's time to go to work."

"A-tten-hut! Get your dogs, get loaded up, and dismissed." We began moving out.

We left the kennels and took off for the truck. I could tell right off that Junior was ready to go to work. I would be last on the truck, but not last off like the first time I walked a post with a new dog.

I had checked the roster and to my good fortune there was Joe on one side and the good Reverend Sergeant Collins on the other side. It would be a great way to get back into the groove of working a dog in good ole Vietnam. Joe would update me on his girlfriend in Qui Nhon and Rev would cheer me up with his homey humor and Biblical insights. Great, life is good.

The truck was taking its time posting the ten dog teams on board. A low ceiling, meaning clouds close to the ground, would mean little light and no planes leaving the ground.

"Hey Moore, how was the trip to Saigon? Did you get in town and see some of the sights?"

I couldn't tell who asked the question coming from the front of the truck, but I hollered back.

"Wasn't a bad trip, and all I can say is we have it so much easier here than those guys. The damn perimeter backs right up to the City of Saigon, in your face like having a city the size of New York for a neighbor. Not cool at all."

"No shit, I heard the city surrounded the base," said an Airman seated next to me. "That must be a nightmare with all the goddamn, 'Halt. Dung Lai!' bullshit."

"Right on, brother, and I'm thankful for bum-fuck nowhere Phu Cat. If it is possible to be in a better place, I don't know where it is."

The truck stopped.

So far, everything has been easy, but it's time for the rubber to meet the road. Here we go, Junior, its changeover time. Will Junior jump my ass? Will the night and the post bring memories back and cause Junior to attack me? We'll see, doesn't make a dang bit of difference. Junior and me are a pair, a couple, married. Till death do us part or I go home next January.

I jumped down from the truck and lifted Junior to the ground, then grabbed my bucket as the truck left me standing on the berm road. I had a surprise waiting for me and it wasn't Charlie. As I sat getting Junior out of his muzzle, I noticed his nose in the air and pointed towards our post.

"Good boy, Junior," I whispered in his ear, "that's a real good boy." I smiled because Junior was going to work without being told.

I sat on my trusty bucket and took my good ole time getting my shit in order before going on my post. I was getting my mind to accept the fact that this was just another night and nothing was waiting to bite me or kill me, I hoped.

I turned my attention to putting the collar on then switching the leash from the choke chain over to the collar. Ears up, nose in the air, something was up. That's when Joe and Prince made their presence known.

"Good boy, Junior." I said. I loved him up, patting and scratching his back side. If he was going to bite, this would have been a good time and with witnesses. Again, nothing happened.

Joe stood in the shadows just off the berm road while I got around to loading my weapon.

I need to start thinking about an overall plan of staying alive and get my head out of my ass. Getting off the truck the first thing you do is lock and load, always. I fucked up already. And how is Junior taking this first night back working the boonies with a new handler who forgot to load his weapon? Does he remember the events from three nights ago? His handler going down wounded? Covering Marcus' body to protect him? Does a dog really give a hoot who handles him? Will he bond with me?

Joe and Prince stepped out of the shadows. "Let's get going, man. You've fucked around here long enough."

"Fuck off. Give me a break." Grabbing my bucket and taking firm control of Junior, I headed onto Kilo 22.

"Me and Rev have already cleared your post. Rev said he would meet up after the coffee truck came by."

"Thanks, Joe. Did you see how Junior alerted on you guys? Fan-fucking-tastic. I was so wrapped up in Junior I had forgotten to load my weapon. Can you believe that shit?"

"Better be glad I wasn't a gook, I could have taken you out so easily. Your dog alerts and there isn't a round in your weapon. Get your head out of your ass, man."

"Okay, I get the picture, so fuck off."

"I'll be right back. Some peaches will taste good right about now." With that, Joe and Prince disappeared into the darkness.

This spot was the perfect hiding place. There were shrubs high enough to conceal me, Junior, and a visitor or two. We had to hide from Charlie ***and*** the towers with their night vision scopes. The bunker guys were no threat to our lives as long as the enemy wasn't around.

Even though Joe and Rev had cleared my post it was my responsibility to do so with my eyes and the eyes, ears, and nose of Junior.

"Do you know, Junior, how powerful you make me?" We took off and headed away from the berm road toward the furthest point I dared to go. If we were lucky, a fence would stop us. If not, we would continue patrolling further and further away from the base. I wanted enough room between me and Charlie that I could get lost, hunker down, and improve my chances of surviving.

Christ, what if I get caught in a crossfire between enemy NVA and the big guns in the bunkers and towers? Everybody trying to kill the same enemy and I'm ten yards away caught in between trying to find a low spot in the zone where VC are pinned down and fighting for their lives. That's life-sucking Vietnam, and not a damn thing to be done about it. Just get down, hug the ground, and pray and get wasted.

I'd be learning Junior's alerts on animals, which were different than alerts on potential enemy. Alerts on bunker guards whispering to each other were still different than a dog's alert on a snake. A rat running through the grass would go unnoticed by a dog, but the poor handler was on his feet with his weapon on full auto, ready to kill something. Older handlers distinguished the different night sounds in a way that the younger guys still had to learn.

I took extra time to clear my post and observe Junior before heading back to "the spot." Joe had not made it back yet so I sat down to cool off. The temperature was at least a hundred degrees and my uniform was already wet from sweat. Junior's tongue was hot and dripping a lot, meaning water was leaving his body. If I didn't get water into Junior, he would overheat and soon be sick. I checked my watch and figured we had been walking for no more than an hour. Pacing our walks would be critical to Junior's good health, mine too, but at least I could sweat all over, the dog couldn't.

Joe and Prince returned and we settled in for some conversation while our dogs cooled down. It was always comforting to have company during the long night. Tonight was special because two of the best were working either side of my post, Joe and Rev.

While Junior enjoyed his water, I opened a can of peaches. Peaches were king over here so early shift guys would take extra, leaving little for the rest of us. We bitched and more c-rations were brought in. Right now, these canned peaches tasted great.

I broke the silence. "I'm glad to be back, the last seventy-two hours have been a whirlwind."

"Hey, I feel real sorry for you," Joe said.

I couldn't help baiting him. "No kidding, Joe, Saigon is a city of lights and the finest women in the whole world."

"You lucky bastard, I would have given my left nut to have visited the place."

I shuddered. "Look, man, I put a dog down in order to visit that place, never forget that."

"Yeah, I know, I know." Joe lit a cigarette, bored. "Tell me about the women, please!"

"Joe, they were out of this world and so many of them. Did you know, man, there are six million people in Saigon, all of them with somewhere to go? I have never seen so many people in all my life. It was kind of creepy, in a way. I enjoyed it, but I would have felt better having some guns on me. I guess I felt naked."

"Like New York, I know. You didn't get laid, did you?"

I shook my head "no."

"I can't believe this shit. You go to the Capital of free Vietnam and you don't get laid? What the f-fuck is wrong with you, man?"

"Don't cry for me, Joe. There was so much eye candy that I just sat eating a steak and taking it all in. Truthfully, I didn't much like Saigon. Too many people who didn't like me and I couldn't understand a blasted word. It made me uncomfortable. My girl-watching would have made you proud, buddy."

Midnight had come and gone and Joe and Prince were back working their post. Junior and I had finished another round of hide and seek and I was thankful for finding nothing. I could hear the coffee truck making slow progress in my direction and knew it would take a while for it to reach me.

I had received several letters while in Saigon and hadn't read the one from my folks, so I opened it to read by the moonlight. I was saddened to learn that my dear Aunt Edith had passed away. As a kid growing up, my best memories of her were the kisses I had to endure every time we visited her and her sister Aunt Rosa, two old maids. My Daddy's sisters sure could cook and they had sugar cookies available almost every time we visited. A nice trade off, cookies for kisses. I would miss her.

Only a few days before my first K-9 party. I hope that nasty typhoon threatening the coastline won't cancel it.

I jumped off my bucket when the rain sort of stopped, and lifted the poncho over my shoulder like a cape so I could cool off. It got incredibly hot in rain gear made of nylon that didn't know the meaning of breathe. I finished relieving myself and took off with Junior in the lead as we went to clear our post again.

So far, I had been lucky not to have any flares going off along the perimeter fence. Dogs knew when there was an enemy snooping around out beyond the depths of our post. We handlers knew it, too, but it was part of the job description so we drank a lot. Drinking lessened the stress of facing yet another night of torrential rains with the usual high winds that brought on rivers of red mud.

There will be a lot to drink at the party. Drugs or drink sure get a lot of people killed over here. Armed Forces TV has a commercial showing a soldier under a banana tree smoking a joint. The next scene, there is only a smoking, bullet-holed helmet left. I enjoy a cold beer after a long night on duty, but I wonder how many of us went on duty, three sheets to the wind.

Eyes and ears were useless on both human and dog and I seriously doubted a dog's sense of smell was any good in this nasty weather. A Typhoon had been beating our coastline for days, was full blown as it moved inland, and night guard was tired. High turbulent winds messed with the scent so much that the dogs were not able to tell which direction it was coming from.

I made a quick walk around my entire post and deemed it clear. Junior never alerted, stopped several times to look and sniff the air. He'd hold that pose as his eyes watered and his red fur moved in circles by fierce winds pounding his face. Lightning flashed and Junior's ear were erect and steady. The head, the ears and the eyes did a ballet of movement. Then Junior lifted a leg for some relief and finally took a dump. I learned to smell the rain.

Nobody but the night guard worked in a Typhoon. Sir Charles stayed home and snuggled.

Junior and I would finish the night in a total downpour. Slipping and sliding, falling down, rain pouring, trying to get up, falling again; my uniform was the color of orange mud, my body also soaking wet which was bad for the crotch and feet. Pick-up time came and the truck filled with weary dogs and handlers, who didn't speak a word, drew up to us. Our dogs tucked themselves between our legs as usual and were also quiet. The trip back to the kennels in the rain was one only a deuce-and-a-half could maneuver through the foot of deep, slippery mud along the berm road.

At the kennels, poor ole Junior was wet to the bone and shaking all over. I gave him a rub down with a dry towel and, of course, a good brushing. This helped to warm and calm him. Before I could turn to leave, his head was on his paws, eyes closed.

"Sleep well, my friend."

Back at the barracks, I had to take a shower to warm up. I was shivering and I had blood on my face from another endless night of sneezing and nose bleeds.

You would think the rain would have washed it away.

I finally lay down and pulled the sheet and a wool blanket up to my chin.

Shit-balls! I'm using a wool blanket in Vietnam to keep warm in eighty degree weather.

The wind continued to blow and the rain came down in even bigger buckets. The sun was already up, but you wouldn't know it from the darkness created by the dying weather.

When I dozed off, my dead friend Jimmy popped up, his dirty face with blood everywhere and a bandage hanging over one eye. My body shuddered and I was as cold as if I were back on post.

How shameful to know how different my war is from the one he fought. He died in the bush, in the filth and heat. Maybe even in the rain.

I lay in my bunk saying a prayer for his soul and all the Marines and soldiers who had it so much tougher than anything we would experience here at the Cat. There were no more faces that night.

Party-time Initiation

When Monday night's duty ended Tuesday morning, all the talk was about the party. It would begin as soon as we returned to the kennels and squared away our dogs, weapons, and gear. I got my share of ribbing.

"What's the big deal? How does this initiation thing work?" I had to yell over the truck flying down the bumping road at thirty miles an hour, with dogs barking and going after other dogs.

"You'll find out soon enough," said one handler.

Another yelled, "You FNGs don't have a need to know."

It didn't take long before several dogs were fighting in the center of the truck and not a single handler paid any attention. I imagined what it would be like if a muzzle came off. With one eye on the dogs fighting, I sat back and enjoyed the wind that cooled my sticky, sweaty face.

The smell of burning meat was in the air when we arrived at the kennels, but the men were ready to get to drinking. The more they drank, the less the steaks mattered. I wandered over to the patio area by the grill to find two big 55-gallon barrels full of iced down beer. I counted twenty steaks cooking away on the grill, but no wonder the meat was burned to a crisp and beyond because I didn't see anybody in charge. There were ten cases of frozen steaks on the ground being used for seats.

Aren't you supposed to defrost meat before cooking it or at least slow cook it? Who's in-charge here?

I found Joe and Vinny standing by the beer barrels and walked over.

"What you drinking this morning?" Vinny spoke with a mischievous smile on his face. "Maybe some Scotch for you?"

"No, hell no," I said. I reached into the barrel for a beer. "You ever going to let me live that down, butt-wipe?"

One morning during a noisy game of poker, I had decided to try Scotch. The music was loud, the smoke was blue and thick around the table. For this non-smoker and others it meant making trips to the out of doors to get a breath of fresh air. I chose the Cutty Sark because of the sailing ship on the label. After a few drinks and some more card playing, I dropped out of the game and headed to the bathroom for a cold shower. I slept hard, didn't hear a thing, but the game continued in my absence. When I awoke, Vinny informed me I had puked on the floor and he had been nice enough to clean up after me. I was embarrassed, and I never drank that shit again.

"No way, scumbag," said Vinny. "Not in this lifetime."

"I see they broke out their best china for the occasion," said Joe. "Hell, the steaks are bigger than the paper plates. I really like the plastic knife and fork, for that special touch."

"Shut up, Joe," Vinny said. He finished his beer. "What are you bitching about?"

"Hey," I said. "Where's the potato salad? You can't have a cook-out without potato salad. At least the beer is free." I grabbed for another cold one.

No sooner had we finished bitching about the food than a truck delivered two large metal containers filled with coleslaw and potato salad. The best part of the delivery was the large tray with chocolate cake. Oh my!

"Cooks wanted to thank you men for the work you're doing." Sergeant Francis set the big tray of cake on the table. "Come and get it, girls, sure looks good. Next time you go through the chow line, try and remember to thank them. They need a pat on the back as much as we do."

By 0830 hours the party was in full swing. At least twenty cases of beer were stacked up on the patio in the shade, and ten more nearby. There was as much of the hard stuff on the table.

"Who is going to drink all this and still walk a dog tonight?" I wondered out loud.

"I'll bet you in the next three hours every bit will be gone." This came from Sergeant Lott who was standing nearby. "I'll put up ten dollars that every beer and everything else will be consumed by ten o'clock."

We all just stared, not saying a word; I thought he was out of his mind. *There are thirty cases of beer for seventy-five men plus all the hard liquor on the table, plus the food.*

"Any takers, you pussies?" Lott asked as he seated himself next to me.

"Unbelievable," Vince said. "No way could all this be consumed in the three hours we have for the festivities."

"Want to bet?" Lott said. "Come on, how about twenty bucks?"

"No, thanks," Vince said. "You've been to more of these parties than we have, so I for one will just have to take you at your word.

"No wonder the Air Force is made up of big time drinkers." I spoke up. "Hell, I'm still trying to keep the beer I drink from ending up on the grass."

I grabbed a half-frozen steak and headed for Junior's kennel. He was already eating from his dish as I entered and put the meat in it. The thought occurred that I shouldn't be this close to an animal eating his main meal for the day.

"Go for it boy, you've earned it, so enjoy."

Junior lay down next to me with a half-eaten bone in his mouth and continued eating. When he had devoured the bone, I gave him some loving up and left him in his kennel so he could rest. I was tired myself, but a party and initiation awaited me.

Rev made it to the party with a Coke in his hand. As I approached the patio area where the grill and drinks were stationed, a few of our group were lying on the ground, either resting before the next round or already passed out. I had never seen so many shit-faced men in all of my short life. I had consumed a six pack by 0900 hours. I wasn't feeling a thing and grabbed another beer.

Back home, ninety per cent of the men protecting this base can't buy beer or the hard stuff, hell, can't vote either.

"Hey, Moore, come on over here and have a steak." Barry Teagan yelled, smiling big from behind the grill.

"How many cases of steak have we eaten?"

"This is the last one, number ten, its ten o'clock and most everything is eaten up or drunk."

"I'm tired," said the Rev, "but I can't leave until I see the initiations. I wouldn't miss this for the world." Rev had a big moonglow smile on his face.

"Rev, what is this initiation all about?"

He laughed, seeing the concern in my eyes.

"You'll not get a thing out of me. Be patient, and it won't hurt for long."

It won't hurt for long?

"Well, it's already ten-thirty and this party has got to be closing down anytime now. Got to get these drunks to bed."

"Jimmie, did you notice the liquor on the table?" said Vinny. "It's almost gone."

"I managed to get several pieces of the cake," Joe said.

"Well, fuck you both very much."

"Well, fuck you, too; you were off with your dog," said Vinny. "By the way, I've been meaning to ask how the new dog worked for you the last couple of nights."

"He's damn good, Vinny. Except for one thing, Junior tried to hump me again last night, and I wasn't about to give in to him."

I finally have a real working dog.

Vinny burst out laughing so hard that others were gathering around to find out what was so funny.

"Junior was humping Jimmie last night, trying to get a hand job!" Vinny burst out laughing and anybody within hearing was laughing too.

"No way do I masturbate a dog. Besides, I save that exercise for myself." I pretended to be embarrassed. "How many of you are jerking off your dogs?"

No one said a thing.

"I know better, guys, most of you are perverts, come on, admit it. This was the second time in two nights he tried humping me, so I pulled him around and punched him right between the eyes."

"Moore, you're damn lucky that dog didn't take a bite out of your ugly face," said Sergeant Francis. "What's your problem? Are you a chicken shit handler, or what? I'll be keeping an eye on you and anybody else who refuses a little ole hand job to his worthy dog."

Everybody was having a good laugh at my expense. It's funny.

A drunk Sergeant Francis announced it was time to get the initiations going. We had managed to consume all the thirty or so cases of beer, ten cases of steaks, all the coleslaw and potato salad, plus the tray of cake.

There were sixty men in various states of intoxication gathered around the barrels of leftover ice cold water, waiting.

"Okay, okay, listen up," said an unsteady Staff Sergeant Francis. "We've fourteen men to get initiated so who is going first?"

There was low rumbling as everybody was looking around at everybody else. No one moved.

I was not going first and still waited to see what they were going to do to us.

It was Russell Puls who came forward first; four men grabbed ahold of Russ, picking him up by the legs, and then turning him upside down, dunked him head first right into one of those barrels of ice cold water. They held him there for a couple of long seconds.

"Okay, okay," said the Sergeant, "Get him out, quick."

Russ landed on his feet with a big ole grin on his face, raised his fist to the sky, and hollered. "Who's next? Nothing to it!"

"Okay, now you guys know what to do, let's get it on," said Francis." We've two barrels and thirteen fucking new guys left to drown. Line up, now, and let's get this over with." The Sergeant was barely able to stand up himself.

The line formed up and the dunking began. It took three or four of the guys to man-handle us into the barrels. We wondered which one of us would drown first. Teagan was a little on the heavy side, so when it was his turn, a bunch of us stepped forward to help get him up and out of the watery grave, with slippery hands grabbing for anything to hold onto.

When it was my turn to be dunked, it was over in a flash and I was grateful for the cool, refreshing water on my body, happy not to have drowned.

Pride and Punishment

Half the time we never knew when our dogs saved our butts from death and accidents, snakes, or plain ole screwing up. One night, Junior stopped me from moving forward, I mean stopped me cold.

"What's going on here, boy?"

He would not let me pass him on the left side. I could back up or pass him on the right side but not the left side.

"Junior, what are you telling me?"

There was something out there. It could not be a snake; if it was, Junior would be out of here in a hurry, pulling me along.

It was dark as hell even though the sky was full of stars. Looking down at the ground, I saw no shadows, just black. I was on my knees with Junior's butt in my face, feeling the pit of my stomach moving to meet my spine.

Phu Cat Air Base was constructed on ground that was confiscated from farmers and their families, along one of the distribution routes of the Ho Chi Minh Trail. It was also an R&R site for Vietcong and the North Vietnamese Army, who dug wells and tunnels for the Air Force to destroy and cover up. All of us dog teams walking perimeter had to be on the lookout for the tunnels and that worried me now. What had Junior found?

I got down on my knees with flashlight in one hand and holding Junior's leash and my CAR-15 in the other. I began to check the ground. Inching forward and using my weapon to poke, I found what had Junior in a tizzy. A hole in the ground that could be either a well or a tunnel. I nervously peeked over the side and down. Junior had not alerted on human or animal, he just sat between me and the hole, about two feet across, to make sure I didn't walk into it. How deep it was I couldn't tell. I got on the radio and called it in.

"How do you miss a two-foot-wide hole in the ground? No telling what was at the bottom." I was talking to Junior but he just looked at me with those sad eyes.

What if it's really part of a tunnel, a new tunnel system? A million gooks could come charging out of that goddamn hole and I would die and not even know what hit me.

"Junior, this sucks."

I'd rather run into a snake in razor grass than a VC with a hatchet.

Junior saved my ass that night and we got the hole covered up. As I sat on my bucket cooling down and hearing Junior lap up his water it dawned on me, *only a nineteen-year-old is dumb enough to do this job. That's why old men don't go to war. I am so thankful for this dog.* My edge.

Kilo One was the easiest post on base. Everybody said so.

One night in April, I sat Junior down on the ground, locked and loaded my CAR-15, and changed my K-9 into a working dog, a killer. It was clear enough why everyone said this post was a piece of cake. First, generators sat about twenty-five yards apart and each had two powerful lights on top facing out towards the perimeter.

Kilo One was the first post and it backed into the Korean Camp. Compared to the Air Force, the Korean troops didn't mess around. Their camp was protected by layers of concertina wire, a minefield, and so much lighting a snake couldn't get close enough to spit.

The post was as flat as a board with nowhere to hide except a few big rocks, so the only saving grace about the post was the unmanned bunker near the main road. It had a roof on it which would be great if it rained.

It didn't look like rain at all but the scene could change at any moment. There were millions of stars overhead to add another layer of lights I didn't need. I called for a radio check and then took that first step.

"Where, oh, where is Charlie tonight?" I hummed to a song from Hee-Haw as I moved out.

I dropped off my bucket by the bunker, put Junior on alert, walked towards the fence line with its lights and minefield, and started the long night of hide and seek. Junior and I cleared Kilo One in record time. Now, we just wanted to get away from the lights and the damn noise of the generators. Junior's sensitive ears were useless and were hurting as much as my ears were.

On returning to the bunker, I again counted it a blessing for its roof if the bottom fell out tonight. I lay out my poncho off to the side of the bunker that was away from the road.

It was still too early to be eating and I wasn't thirsty. *The guys were right; this will be a piece of cake.*

Right behind the bunker, the flow of traffic on the main base road that ran to my rear from south to north was a small distraction. I listened for the nightly run-ups of the big F-4 jet engines to begin. Lucky for the both of us, they were far enough away where the sound couldn't hurt Junior's ears.

I took a pair of drumsticks out of my bucket and started to pound on the hard ground. It wasn't loud and it helped me focus on the real estate out in front of me. I practiced para-diddles and rolls and flams and a whole lot of other things to help pass the long night.

Around 0200 hours when I felt the urge to relieve myself, I grabbed my roll of toilet paper and moved away from the area of my poncho to a good distance from the bunker. There was an unwritten code amongst us handlers not to eat, pee or dump where we would be hiding. Most followed the code but a few didn't.

The only screw up I made that night was forgetting to have the leash on my wrist. As I stood up and began pulling up and buttoning my pants, I heard a scream from behind me.

What the hell?

Then I heard growls and barking. It was Junior and he was not attached to my wrist. I'd been so involved discharging things that I'd forgotten about him.

"Get this fucking dog off us!" Someone was screaming.

I have no idea who's on my post.

I grabbed my weapon and ran for the bunker. "Junior, out boy! Out, Junior. Sit."

Junior was in the sit position but barking at two men in the bunker. I couldn't help but let out a little snicker as I grabbed the leash and pulled Junior away from the men. I really couldn't see a thing. The headlights of their jeep lit up the place, but I was blocking the light on the two men.

I pulled Junior to me and praised him up.

"Good boy, Junior, good boy." Damn, I was excited as hell, and so was Junior. I just couldn't help having a big shit-eating grin on my face.

"Damn good job, Junior. Good boy."

I got carried away, but Junior knew he had done a good job so I wasn't about to correct him for doing what he was trained to do.

"What do you mean by turning that dog loose on us?" said one of the men. They were still hiding in the bunker.

"First off, I didn't turn my dog loose on you, and second, you're on my post uninvited. You're damn lucky not to be bleeding right now or shot." I was mad as hell and continued to praise Junior.

I still didn't know who these guys were and was thinking of turning Junior loose on them again, but they were plenty scared and pissed at me. Junior did not like these two at all.

"Come out and identify yourselves, slow and easy. I don't want Junior to get all riled up again."

It turned out they were officers. I let Junior bark, and to his credit, he stayed in the sit position. The two a-holes did not budge from the bunker.

"What are you two doing on my post? I received no radio call that I would be having visitors tonight. What is going on? Rightfully, I could have let my dog have a piece of you."

I was screaming at these two idiots, Junior was barking his head off, and the noise from the generator was getting louder and louder.

"Airman, I'm going to have your ass for this." One of the Lieutenants spat at me.

Junior broke for him and I stopped him at the end of his leash. Pulling Junior back, I began praising him up. This was one pissed off Lieutenant and Junior did not like the tone of his voice or his body language at all, so Junior charged at him.

"Airman, you get control of that goddamn dog right now. I'm going to bust your ass for this. You're going to jail." He spat at me again.

I got Junior back in the heel position. "Good boy, Junior, good boy," I said. Getting a chance to see Junior in action was priceless.

They both were giving me a tongue-lashing and it was all I could do to keep Junior from going after them again.

I pulled Junior out of their path and they stormed off to mount their jeep. Before they departed, they pulled beside me. "Airman, what is your name and rank?"

"Airman First Class Moore," I said.

"What post is this?" The Lieutenant asked with a snarl.

"Kilo One, sir."

The Lieutenant gunned the gas and was out of there.

I just stood there holding my belt and wondered what had just happened. *Am I going to jail?*

"Going to jail?" I said. "For doing my job, our job? Good boy, Junior, good boy."

I spent the next four hours wondering when the hammer would fall. By the time the relief truck picked me up, I had my story all worked out in my mind and felt there was nothing to worry about.

I put Junior in his kennel and went to the K-9 office.

"Moore, First Shirt wants to see you at 0800 in his office," said Staff Sergeant Francis.

"Yes, Sergeant."

I put up my radio in the rack, dropped the battery into the charger and headed for the weapons area to clean my weapon. It was a little after 0700 and I figured I would just hang around the kennel until my meeting with the First Shirt.

The dreaded hour arrived and I knocked on the door.

"Come on in." Sergeant Francis was smirking.

Sergeant Johnson was behind his desk displaying an unhappy face. I stood at attention in front of the Sergeant.

"Airman Moore, reporting as ordered, sir."

"Moore, I've got two very pissed Lieutenants wanting me to write you up on an Article 15 and have you Court Marshalled for sleeping on duty. You damn well better have a better story than theirs."

"Sergeant, none of this would have happened if they had called over the radio."

"Moore, they're saying they tried to make radio contact with you."

"Tried, sir, and the assholes came on my post anyway. Hell, they're lucky I didn't shoot the dumb bastards."

"At ease, Moore. These are serious charges and I'm the only one who can save your ass. Do you understand that?"

"Yes, sir, I think I understand."

"Just tell me what the hell happened out there."

"Sergeant, it's real simple, I was taking a shit, I heard my dog bark and growl and then these two men were screaming to get this fucking dog off their asses. I was pulling my pants up, trying to run as I grabbed my weapon and circled around to the front of the bunker. Junior had these guys pinned inside the bunker. So, I'm trying to get Junior under control and hold my pants up and not drop my weapon and figure out who is on my post unannounced."

The Sergeant was waiting for me to finish. "Moore, are you telling me everything? These men say you were sleeping on duty."

"I was not sleeping, I was taking a crap. I left the pile of shit uncovered. Besides, they had the Jeep's headlights on and shining right on me. They saw me holding my pants up. I was not sleeping on duty."

"Master Sergeant! I was doing my job, real simple. This is officer fuck-up and poor airmen gets to pay."

Sergeant Johnson glanced over at Sergeant Francis who was carrying the biggest grin, and boy was I happy to see that. There came a little hint of a chuckle from the First Shirt and I began to relax a little.

"Okay, Moore, I believe you, but these Lieutenants are pissed because you made them look silly."

"Looking back now, I may have praised Junior a little too much. But how often do we really get to see our dogs in action? Junior did his job, but I fucked up not having the leash on my arm. I didn't want shit on the leash or to get that mess on my hands. Sergeant, if these people had made contact with me, this would not have happened."

"You hurt their egos and they want some satisfaction, so here is what we are going to do. Starting tomorrow, you are to report here at 0100 hours for the next ninety days for two hours of KP."

"Ninety days! Are you shitting me? Sergeant, I did my job, they made no radio contact. They came down on my post like a bunch of school girls and got nailed, and I have to pay?"

"I agree with you, Moore, and I believe your story. Punishing you is not something I take lightly, but these two Lieutenants that you supremely embarrassed want blood. I will convince them that ninety days of KP is enough punishment. And if I have to get down and dirty, I'll remind them they made no radio contact before entering a secured K-9 post in the middle of the night."

"Yes, Sergeant, is that all?"

"Yes, Airman. You're dismissed."

All the guys playing cards around the table got a good laugh at my predicament and accused me of fucking off and sleeping on post. They said my ass deserved to be in jail, or better yet, in front of a firing squad.

"Eat shit and die, scumbags."

Between listening to the taunts coming from the card game and thinking about getting even, I couldn't sleep. I would show those sons of bitches. I got up, dressed, and headed for the door with the intentions of taking my anger out on some breakfast.

Angry or not, I served my KP duty. Getting ready to go to work the next night after being hit was tough duty, and thirty days of quiet, peaceful nights later, we were still nervous wrecks with anticipation. After a while we became almost zombie-like, careless, to the point of numbness from the slow drain of energy. How long will liquor soothe our nerves?

No wonder, one night months later during guard mount, a handler came staggering toward us wearing civilian clothes, carrying an unloaded weapon in his left hand, his dog with no muzzle on, on the other hand.

The only thing we could do short of an Article 15 was to put his drunk ass to bed and cover for him. We put him to bed. Next day he was ribbed mercilessly for messing up the rotation. That's called taking care of your own.

One morning, as the relief truck driver made his rounds, he spotted a dog minus his handler waiting at the pickup point. To the handlers and dogs already mounted up, it was funny, but the driver had a decision to make: "Do I leave him or go and find his sleeping ass?"

The driver got the dog in tow, found the handler, and chewed his ass on the walk back to the truck. Not another word was spoken about it other than a few laughs over card games for the next week or two. The lifers never knew.

There were plenty of us who puked before taking that first unknown step into the darkness. I was one of them, but it didn't happen often. Puking got boring after a while. There was no choice but to man up every night.

Chapter Nine

Rok and Roll

Phu Cat Air Base, Vietnam – June 1969

US Headline News

Stonewall Riots in New York City Mark Start of US Modern Gay Rights Movement.

Judy Garland Dies of Drug Overdose in her London Home.

We had been hit twice since my arrival, but had no new wounded and just a few enemy KIA. It was hard to stay sharp when there was so little action. We all got sleepy on post sometimes.

One night in June, I happened to have the post next to Joe and met him as he dismounted the truck. He had something on his mind. I asked if he went into town today.

"Yeah," he said. "How is my post looking, find any gooks on it?"

"Joe, your post is all clear, I walked it several times."

"Thanks, pal."

Joe looked worried.

"Buddy, it's okay, I cleared it, and there are no motherfucking gooks on either of our posts."

"Who's working the post next to me?"

"Don't know him."

I could tell Joe was nervous about something. So was I. I didn't like the late shift.

"Hey, Joe, no rain for a while, can you dig it?"

"Damn right, I hate the fucking rain."

"Joe, my man, the rain is a great equalizer. Charlie is in his nice, warm bed listening to the rain on the roof tonight. He is not out here sneaking around, freaking all of us out, upsetting the dogs. Right about now his wife joins him for a little fucky-sucky. Charlie is no fool; he lets us deal with the rain."I was hoping to get a little laugh out of him.

"Dumbass, it's sucky-fucky, not fucky-sucky."

"Up yours," I said. "You ready, Joe?"

"I'm ready. I'm just m-messing with you, man, okay?"

He gathered his bucket, weapon, and dog and headed into the unknown, hoping I was right that Charlie was sleeping in the arms of his wife.

Just getting down the slight hill without falling was quite an accomplishment. It was slicker than snot, and when the lightning flashed and the terrain opened up before us, we knew it was going to be a long, cold, miserable night.

"Best we settle in and catch a break before monsoon starts up again," I said.

We alerted our two dogs to "watch 'em, boy" and waited for it to darken as a large cloud covered the moon. We couldn't afford to be all lit up. In fact, being flat on the ground was the best hiding place. Walking around on post was broadcasting to a sniper where we were and the moonlight made it easier.

We're nothing but cannon fodder, the new guys get every shitty post and duty. It will be that way for a long time, six months already.

As we refreshed ourselves and both dogs, I asked Joe how his girlfriend was doing.

A big smile came to his face. "I believe I've found love. Can't believe how l-lucky I am."

"Well, when are you going back?" I asked.

"Just as soon as I get off work in the morning, I'm headed for Qui Nhon."

"Joe, you're going to disappear one of these days."

"No kidding, today while I was making love to Vera, some of those Cowboys crashed through the door and four men dressed in all black suddenly surrounded our bed. Their leader put a gun to my head and a knife to my naked balls. The other boys took my watch and money and backed out of the room while Vera was still trying to knock out one of them with her tiny fists."

"Did she know these guys?"

"Yes, and that's when I left. She begged me to come back and said she was sorry. Outside the door I could still hear Vera crying and I wanted to go back in, but I knew I'd better leave."

He was crazy for Vera, but I wanted him to think. "Hey, did you hear about the lifer who was shipping out in a few days and had to have one more piece of ass in downtown Qui Nhon?"

"No, what about it?"

"The whore had a razor blade in her cunt and chopped his pecker up pretty good. He won't be going home for a while. Can't have his wife sucking on a mutilated cock, now can you?"

We really didn't give a shit about some wasted lifer getting his manhood chewed up. Joe and I had a good laugh.

"I know you're careful, but damn," I said. "You're taking your life in your hands. You're mighty calm about a knife meeting your balls."

"Jimmie, here's how I do it, the same way this dummy lifer did it. I go through the gate that connects us with the ROK camp, walk right out their main gate, and hitchhike to town, nothing to it," Joe said. "I've been doing it for m-months with no one the wiser except you and Vinny."

"Joe, you got guts and big balls, is all I can say."

On the ride back to Phu Cat, Joe couldn't get it out of his mind that Vera knew those bastards. He realized he was alive because Vera knew his would-be attackers.

"Maybe I could get my watch back after all. Would they really have hurt me? Did she really stop them from cutting me? It makes me want to scream."

I nodded and went to my bunk, listening to the news on television and trying to decide whether to get up and eat some chow.

The TV announcer says the population of the United States just hit two hundred million. Next door to Vietnam is China with almost a billion people.

I enjoyed the big sky and star watching in Vietnam. Almost every night there were stars streaking across the sky. One night, I was lying on the ground watching for falling stars when the ground began to vibrate like an earthquake. It seemed to go on and on forever before the ground stopped shaking. Junior gave me a quizzical look.

What's it like to be aboard a B-52 bomber? I can imagine sitting at the controls, lining the plane up for a bombing run, feeling the bomber go light as the bombs leave the bomb bay and start their journey to deface the earth. Who cares who dies at the end of the journey as long as the effort slows down the ants bringing supplies down the Ho Chi Minh Trail? I hear they fly their missions from Guam. That's a long time to go without a bathroom break.

Spooky, the gunship, was another sign of American air power in the air most nights, and I'd watched the impressive show many times before. We had several of the C-130 Hercules here at the Cat with mini-guns mounted inside. They put out so much fire power that at a distance it looked like a scene from *War of the Worlds*. That ray of red death killed a lot of Vietnamese, but saved a lot of American lives.

If letters could spell the sound of the firing guns, it would be like this: "BBbaharrrooooommmm," for about five to ten seconds.

I loved the sounds of war.

Dark Dreams

My dream was of a face. This face took the place of Lillian in my dreams and that face was mean. It came more now that I had passed the six month point and was marking the days off until I could return to the world.

I had seen the face for real, and I remembered where I had seen it, on my first visit to Qui Nhon. The next time I about shit a brick and the hair on the back of my neck stood up when I saw him in the barber shop on base. My heart raced, and I wondered if he remembered me.

Damn, was I fucking his sister? No. Shit, it was his daughter!

Our eyes met. "Airman, how you want hair cut?" said the barber.

"Short," I said.

Does this fucker remember me or not? I think he does. Is he VC? Will he find a way to cut my throat? Thank God, I shave myself.

I sat in the chair trying not to make eye contact with him by looking in the mirror to the front. He was the same man who had watched me having sex with – who knows? In Vietnam, everything was out to kill you. And having sex was no different.

The closer I got to leaving Vietnam, the more I was seeing this face in my dreams. The more I saw the face, the more it was distorted. The thin, straggly hair hanging at the corners of his lips covered a twisted, foul grin. He watched me with sunken eyes, black and set so deep into the skull as to be formless.

Why is this man in my head?

Since I had bought a small fan and mounted it at the head of my bunk, I possessed the secret for sleeping in the heat of the day. Soon I was dreaming the sweet dreams of Lillian that brought comfort to my sleep.

I was awakened by the sounds of the television program called *Hee-Haw*. I knew that it was six o'clock and the first shift would be headed for work in an hour. I got dressed while Roy Clark on the TV sang his hit song "Yesterday When I Was Young."

At guard mount, Staff Sergeant Francis gave our briefing. "We had a dog bitten by a cobra last night, which means one less dog patrolling tonight. It caused the dog's head to swell up twice as big, but he'll be back on the job in three or four days."

I hate walking in the dark on ground shared with snakes galore. Hope Junior's on the ball tonight.

"On the north side of the base there's been movement every night. When you work the northeast portion of the perimeter, you have to be at your best."

Already, Sergeant Francis was huffing and puffing from the exertion of his report, so he paused for a moment before continuing. "Keep your heads; it makes for a less nervous trigger finger."

The third hit by the VC happened in July. As usual, their sappers took on the first line of defense, K-9, and lost. I was on Kilo 40, well away from the action. Three times in a row I had been lucky.

There were no casualties on our side, but the VC suffered one wounded in action and one killed in action who turned out to be a barber on the base. I knew which one, and I stopped seeing his ugly face in my dreams.

◊ ◊ ◊

July 1969

US Headline News

Apollo 11 Astronauts Return from First Successful Moon Landing.

Vietnam War: First US Troop Withdrawals Made.

Rev and I were sitting on our buckets taking a break on the two kilos located at the extreme eastern end of the base. I was again on the same post as my first night walking Duke II. Unlike that foggy night, tonight was a breezy, hot one and nothing but millions of stars to light the way.

Rev Collins had already spent a year-and-a-half in country before my group arrived and planned to finish his four years in the Air Force right here at the Cat. At the end of his first twelve-month tour he'd taken a thirty-days leave, and that's when he returned home for a month to visit with his family and friends back in South Carolina. Then back to the Cat for another 12-month tour. Nuts to that.

He also preached during his time back home. Rev truly lived for the Lord and loved to talk about Jesus Christ. When guys would poke fun at him behind his back, I found myself standing up for him. Everybody at one time or the other tried to lure him into a card game or tempt him to share a beer with us, but he would have nothing to do with it. He was a straight shooter.

The Sergeant and I sat eating peaches with Rev doing most of the talking. "I heard bits and pieces about your run in with the lieutenants on your post. What was that all about, and why didn't you kill them?"

"You're serious about killing them? Would you really have shot two Officers on your post?"

"Damn right, I would have. There was no radio contact at all?" said Rev. "Are you sure about that?"

"Of course, I'm sure about it, I could have shot the idiots," I said. "Besides, I was trying to pull up my pants from taking a dump, and good ole Junior had them pinned in the corner of the bunker with them screaming at the top of their lungs."

Rev shook his head in disgust.

"You should have heard them, Rev. 'Get this fucking dog off us!'"

"Jimmie, you really have to watch your mouth."

"Sorry, Rev, but it's hard to tell the story without getting upset and anyway, they said it. Once Junior was under my control I started praising and loving him up and I think that's what really pissed them off. They went ballistic, and now I'm pulling ninety days of extra duty I shouldn't be doing."

"No radio contact," Rev said. "They were very stupid. Any time someone walks on your post he is a potential enemy. I would have shot the fools. Now, let me tell you a story that happened right where you are now."

"I'm all ears."

"I was sitting on my helmet eating peaches. In the same instant, I heard gunfire and saw the bullets hit the ground not more than two feet in front of me."

"No way, you're kidding, man." I waited for the Rev to continue, after a pause stuffing a peach in his mouth that lasted forever, Rev finally spoke.

"Jimmie, I kid you not. I didn't move or stop eating my peaches."

"Who was shooting at you?" My excitement was building. "Come on, Rev, cut the crap and tell me what happened."

"It was the ROK who took that shot at me and I knew it."

"You mean the Korean Army guys who are supposed to be guarding their water well out front of us?"

"The very same and I'm telling the truth, as God is my witness. They have Starlight Scopes and can see you and me right now."

"Rev, forgive me, but I would have shit all over myself. I would have been on the radio and raising all kinds of hell."

"No big deal. I knew what was going on and I waved at them. I could hear them laughing."

"Crazy, stupid K-9," I said.

"You got it. Bottom line, these Tiger boys have respect for us because we do a crazy, stupid job."

"I guess so, but couldn't they just as easily yell out a hello?"

"Probably, but they were just having some fun. And the fact that it didn't bother me, I'm sure they got off on that."

"So you just waved and kept eating your peaches."

"Hey, I'm telling you this story for a reason. Will you stop and think about what's really going on around you?"

I could tell Rev was getting serious by his tone.

"It could happen right now, one shot fired, lands right in front of me."

"I'd freak out."

"Jimmie, don't over react. Someone might be sending you a greeting and a warning that maybe you need to hide, take cover."

Rev picked up his weapon and pointed it at the stars. "I want to kill the enemy, even if it's just one VC or NVA, I don't care. I know that sounds strange coming from a pastor and a Godly man. I have no love for these people at all and I want to kill at least one gook before leaving this place. I've prayed about it, believe that, I have, and I'm not ashamed about it. Please Lord, just one before I leave."

"I can dig it." I had my own surreal hate for this people and their country.

How does a man of God dream of killing gooks?

Rev got up and left. I couldn't help but think about the Koreans scoping me out as I finished my peaches. I lifted my hand and waved in the direction of the small camp, and to my surprise, I could hear laughing.

"I'll be damned, too damn spooky."

If only I could see past all this dark. A Starlight Scope would make life easier and safer. The towers have them and I have yet to hear of any VC being spotted by them using a scope. But what do I know?

We Can't Do That, Can We?

I had finished eighty-five of the ninety days of KP at the kennels. Feeding, washing out their kennels, and filling their buckets with fresh water allowed the dogs to get to know and accept me.

While I was pulling my final few days of KP with all my doggie friends, I stepped off the bus one morning and heard a racket.

I entered the K-9 office.

"What is going on with the dogs, what has them so riled up?" I asked one of the guys cleaning a weapon.

"We have a POW locked up in a kennel and just about every dog wants a piece of him. The Koreans brought him in."

"No shit? I've got to check this out."

Dead center of the kennels I found the gook tied up and blindfolded. His feet were tied to the front of his cage and his arms and hands were stretched out and tied to the rear of the kennel. All he had on was underwear. It had to be in the high nineties with no wind but this poor fellow was shaking all over. It was the dogs, of course, and he was frightened.

I would try and calm the place down.

"Shut up." That got their attention.

"Out." The dogs wanted to attack this poor VC so it was worth a try.

Still silent, this may work.

"Down." I was hoping to get the idea of sleepy time in their minds. Still silent. I couldn't believe it but the dogs settled right down and later, as I refreshed their water, most were asleep. *Sweet dreams, boys.*

After cleaning all the kennels as best I could, I returned to the office to get the scoop on our visitor.

"ROK brought him in a couple hours ago. We don't know if he's VC or just in trouble. For my money," said the two-striper sitting behind the desk, "I bet he's VC. Hell, isn't everyone VC here?"

It wasn't much of a joke to me.

The Vietnamese are in our mess halls cooking most of the food and serving it. No telling how much spit or shit goes into our food. Gooks can be found everywhere from the flight line to the BX barbershop to washing our clothes and cleaning up after us.

"After the ROK Army gets done talking nice to him maybe they'll want a dog to get in his face. I would love to see Junior take a bite out of his ass."

"We can't do that, can we?" said the two-striper.

I didn't answer because I was already on my way to check in on the POW. He was fast asleep, as were the dogs. Only a few dogs were on their feet watching me as I moved around.

"Down, boy," I whispered. "Go to sleep. Good boy."

Some Airmen had to clean kennels of shit before they would go in and retrieve their dogs. Joe was one of them. His dog Prince was famous for getting so excited to hear the trucks come in that he sent dog shit flying all over the walls of the kennel, even on the ceiling's light fixture.

It only took a couple times for Joe to figure he would have to clean out the kennels and wash down Prince every night, and he hated it. Even with all the precautions, Joe would still come to guard mount with fresh shit on him. We would tease him for smelling, calling him 'stinky butt'."

My dog Junior held it, and I made sure he got to relieve himself before getting on the truck. Most of the dogs would crap in a corner and never go near it, or not shit at all.

Airman Mike Green (not his real name) was a close neighbor of Vinny and Joe from Jersey City. He was tall, at least several inches over my head, with brown eyes and hair, a swarthy complexion, and I hated the sight of his shirtless chest being kind-of deformed, sunk in. I wondered why Mike didn't hang out with Vinny and Joe.

I had spent several nights visiting with him and found he was easy going, took care of his dog, and as far as anybody knew, he did his job. Our conversations on post during the long nights of boredom exposed things that would go unsaid back in the World, like when he told me how rough it was growing up in Jersey City.

"I was beat up a lot by gangs," said Mike.

"The only thing I know about gangs is from the movie, *West Side Story*."

Mike continued, "Then I'd go home and get beat up by my father. An unemployable drunk."

"My father whipped me a lot, but I guess I was getting my due. Nah! Not really." I said. I couldn't relate when Mike explained the beatings he got from his father were like the ones he would get on the school grounds by boys bigger and stronger than he was.

Mike bunked in a different section of the barracks, and the guys who occupied his area got fed up with him not bathing, getting his uniform washed, nor changing his sheets in months. Vinny kept tabs on Mike, and one night on post he told me this tale about what happened when Mike's body odor drove a few of the guys to the First Sergeant to ask for help.

"Give him a bath with a couple stiff brushes," said Master Sergeant Johnson. "This will cure him. Then explain to him that a mama-san will wash his laundry, all of it, on the cheap. Also, take him down to the central laundry and show him where he can get clean sheets once a week."

His bunk mates chickened out on the cleaning with brushes. Instead they took him by the hands and walked him through the suggestions the First Shirt had offered.

Mike still didn't shower.

The Airmen returned to the Sergeant to ask permission to wash Mike. "We don't want to hurt the man, just clean him up and teach him a lesson," said their leader.

"Oh, you'll want washcloths, I guess? Use the brushes, but be gentle, or as gentle as it takes to get the message across to him: Clean your sorry ass up or we'll be doing this again and we won't be so gentle about it next time."

The Airmen nodded agreement.

The First Sergeant continued. "After you scrub him, I will personally inform Airman Green that it is his responsibility to clean the whole room from top to bottom and every corner. I will give him three days to get the job done. If it's not done to my satisfaction, then Airman Green will face disciplinary action with an Article 15 and ninety days of KP at the kennels, and cleaning the whole barracks building top to bottom, end to end."

Another handler spoke up. "Sergeant, we think the embarrassment of seven or eight men stripping him and throwing him in the shower to scrub him down with these disgusting old brushes will do the trick."

"What happened next?" I said.

"The next day around lunch time," Vinny said, "seven handlers grabbed Mike and dragged him to the shower. They ran several of the mama-sans out and stripped Mike, even of his underwear."

"Naked?" I said. "Oh, my gosh."

"Mike was screaming his dumbass head off, but once he was naked we went to work on him taking turns scrubbing him.

"We? You, too, Vinny? Poor guy."

"My ass! The guy is one nasty fuck and he got off easy. We scrubbed him until he had blood oozing from spots on his body. It took almost an hour and we finished up by cleaning his ears, face and his nasty hair."

Vinny's mouth formed a contemptuous scowl. "How do people let themselves go like that?"

"I don't know," I said, "but I never looked at him as being dirty."

"We are Air Force, not grunts who have no choices about cleanness and go for days or weeks in the boonies. We have hot showers, for God sake."

"Well, it's been a week. How's he doing?"

"I think we got through," said Vinny. "He cleaned his section and did a great job, it's spotless. But you better believe we will be keeping an eye on him."

"That's great," I said, "now let me tell you a really gross story about a woman in Jacksonville who recently died. When they cut her open and looked around, they couldn't find a cause of death. Do you want to know how she croaked?"

"Can't wait, I'm all ears," said Vinny.

"This woman had big hair she wore on top of her head in something called a bee-hive. They started to remove the hair and roaches came out of everywhere. Even the medical examiner was grossed out."

"No fucking way, that's a bullshit story if I ever heard one."

"I'm not shitting you. This woman sprayed her hair, and sealed it, and never washed it. She had no idea that roaches were eating into her brain."

"Too bad we couldn't have gotten ahold of her and cleaned her up, maybe introduce her to Green."

Vinny and I laughed so loud that everybody in Vietnam knew where K-9 handlers were fucking off. We decided to return to our own posts and check for roaches.

Tigers in the Bush

I was fifteen days from serving ten months in Vietnam, half way through October, and I was starting to think short-timer sounded good. We hadn't been hit since July, and despite being warned not to let our guard down, I had. At times, I forgot to load my weapon. A few times I slept on post, especially if it was raining. It was stupid to sleep on post and most didn't.

One morning, Vinny, Joe, Barry and several others were playing cards when one of the day crew from the kennels came in looking for volunteers. A three-striper none of us knew well barged in.

"Hey, guys, the First Sergeant is looking for two volunteers to go on a special mission with the ROK Army."

Joe ignored the question. "How'd you get the day shift?"

"Lots of ass kissing, Joe. What do you think?" said Barry.

"Boys, I get the day duty cause I been here over two years. I'm the one that cleans out the kennels and feeds your dogs every day. I'm also on call every night in case we get hit. I'm one of the guys who comes to the rescue. So, do any of you wish to volunteer?"

I raised my hand. "I'll go. Do I get the night off?"

"No, Moore, you don't get the night off but your name will move up the roster. Let's go, the truck's waiting on us."

I was disappointed that Junior would not be used on this mission with the Koreans. Instead, I would carry extra ammo and water in my bucket. I was taken to the terminal building for a ride on an HH-43. This bird to me looked like an upside-down egg beater with blades set next to each other but tilted so as not to hit one another. Ugly things.

We mounted up, buckled in, then the helicopter called Pedro revved up and took off.

If the chopper doesn't kill me, and if I survive the day with the Korean Army guys … I volunteered for this?

What a ride! This bird shook, rattled and rolled, and if not for the belts holding us in, we thought for sure we were going to fall out of the damn thing. Brad, the other handler, and I were bug-eyed.

At least he gets to have his dog along.

After about a thirty-minute ride we began to lose altitude. I still couldn't see anything of the Koreans we were supposed to meet.

The rear end of the Pedro turned 180 degrees, and a small village of about twenty or twenty-five structures appeared, spread out over a football field-size area. It was a farming village built with the typical materials of the rural areas in Vietnam: bamboo, stucco, odd pieces of lumber, plywood, and tin. The village was surrounded by rice paddies full of water that had not been planted since the new air base had been built back in 1966. Now it was a free-fire zone where anybody or anything caught inside was considered VC and could be killed on sight.

No telling how many villages like this one were uprooted and the people displaced.

I had been briefed on the way. "If you catch them in the free-fire zone, you must give these uninformed villagers several warnings. Say, 'Halt. Dung lai,' so they know what you mean."

Just one more rule of engagement some fat-assed drunk of an officer put together. I want to kill the enemy, not give him the benefit of the doubt.

On the ground the trees and hedge rows were big enough to hide a lot of trouble. I understood why they wanted to try out one of our Sentry Dogs' powerful noses. Brad changed over his dog Silver while I locked and loaded, then got out my canteen and made ready with some water for Silver.

Brad loaded his CAR-15 and we stood observing the relaxed ROK Tiger Division troops (also know as the Capital Guard) who were smoking and drinking water. They had a smile for us as we made eye contact with them. These men were the same ones who protected the Cat from attack and were supposed to be a tough outfit.

I recognized one who had been our karate instructor from way back when.

The Pedro took off for the return trip to the Cat. A Korean Lieutenant gave us our instructions. "This is a search and destroy mission. We need to clear a village and we hope that a K-9 will help without getting any of my people hurt."

Brad looked at me and spoke in a whisper. "Stay close."

We moved out with Brad in front taking the point and Silver leading him. I followed up with extra ammo ready to give out if we were attacked. I carried my bucket with more ammo and hoped the Koreans had more ammo if we ran out.

"Moore, keep your hand on my belt so I'll know you're there. You hear me, Moore?"

Brad and I were both trembling and breathing hard before we started.

"I've got your back, Brad. You ready?"

I wasn't feeling as brave as I hoped, but Brad knew I was right behind him.

"Watch 'em, boy. Silver, watch 'em." Brad whispered in Silver's ear and we moved out.

I never took my hand from Brad's web belt as we walked along the well-worn wooden side of an old building that may have been a barn. It had a lot of bullet holes in it. The hair went up on the back of my neck.

Damn, I've put myself right in the middle of the real war.

The sun was blinding hot and sweat was already blurring my vision. It was then that I realized how serious the heat was for the dog and we needed to be careful. We had already been in the field forty minutes.

Silver came closer to the corner of the big barn and I felt Brad stiffen.

Has Silver alerted?

Brad stopped and dropped to a knee, so I did likewise. I could see over the top of his head that indeed Silver alerted to something. Was it human, animal or something else? Silver was in no hurry to move forward.

Brad turned and shrugged his shoulders. I looked back at the Lieutenant kneeling behind me and pointed with my thumb to the interior of the building, shrugging. We just couldn't be sure.

Looking through the many bullets holes into the abandoned buildings made the job go easier. Silver checked ten or eleven buildings of various sizes, right down to small hooches made of bamboo. He would stick his nose through a door and, on a signal from Brad, several troops would enter the building and clear it.

It took time prying underneath flooring and in roofs for possible hiding places for weapons, ammo, food, and people. Once the building was cleared the dog was given the chance to sniff out the flooring before we moved on to the next. So far not a single gun was fired. Our ROK friends took all the risk. All we had to do was let Silver do what he was trained for, hunting and killing.

We had cleared almost two-thirds of the structures when Brad looked back at me. "Jimmie, I've got to get some water in Silver. I've already overworked him."

I turned to the Lieutenant, pointed to my canteen and to Silver, and he got the message. The Lieutenant had us pull back to a rest area, and then troops of the Korean Army came out of the woods lighting up smokes and going for water. The grunts were standing around looking at Silver as he drank water. You could tell in their smiles they were pleased how the dog team had worked out so far.

A Soldier walked over to me and handed me a large chunk of watermelon. I took the piece of pink fruit, bit into the center part and came out with seeds and melon juice all over me. The grunts broke out into a laugh.

Brad took me aside. "Moore, you know what they say about eating or drinking anything not from the base."

"Yes, I do, but wait until you get a taste of this watermelon," I said. I took another big bite "It's really good, try it."

One of the ROK Tigers approached Brad holding out a piece of the melon for him. Before Brad's shouts could stop him, Silver lunged at the oncoming warrior. Brad pulled hard on the leash, just in time to save the dummy from a nasty attack that was headed for his face. The soldier fell back on his ass in shocked disbelief, with a gaping month.

The ROK troops couldn't contain themselves. They gave the target of the attack a hard time.

I think they're falling in love with Silver.

"What is it about these dogs that make them like going for the face?" said Brad.

"Damn if I know, but it sure was funny."

The sun was straight up in a cloudless sky when we started back to work. The grunts spread out with most of them disappearing. Almost an hour went by, and we had most of the remaining buildings cleared. Silver alerted, and the Korean Soldiers found cover or hit the ground, expecting the worst after being too long in the bush. I hunkered down myself behind Brad and I laid my hand on his back, just to let him know I was there.

Brad made eye contact and shook his head.

"Brad, you're not sure what the alert is for?" A drop of sweat entered my eye and burned.

"I don't know what he is alerting on but I don't think it's a man. More than likely, it is nothing."

Again, with a shrug of my shoulders, I told the Lieutenant we weren't sure whether anything was in the building.

He got on his radio, and at the same time motioned for us to pull back to the rear.

The ROK soldiers were in motion all around us. Several moved close to us as if to protect us.

"Get down, get down, GI." They whispered and made a hand signal that meant to get the fuck down.

I was already on my belly and watched as Brad put Silver in the down position between him and several bamboo trees.

"B O O M!"

One of his men had thrown a grenade into the building. Next,

Soldiers were up and running towards the front and entered the smoke-filled building to clear it.

The ROK Lieutenant returned with a smile on his face and holding several objects in his hands. "Good job, very good job. Thank you very much," he said.

In his broken English, he said that the VC had been there today, but pulled out early, leaving behind a small quantity of ammo and some cooked rice that still held some heat.

The Soldiers were cheering. They had big ole shit eating grins from ear to ear.

There is nothing cheerful about us just missing the possibility of being ambushed. Get me out of here.

"If I had known what I was getting into, no way would I have volunteered," said Brad.

"This dog proved the worth of all our sentry dogs. Good work, Silver," I said.

"Yes, good boy, Silver," Brad said. He gave Silver lots of pats and loving up to show his appreciation.

The men of the Korean Tigers from the Capital Division seemed to relax, but not really. Our ride back was on its way and they all said the mission was a success. The Korean Army Lieutenant shook my hand, then saluted Brad and Silver. Silver didn't care because Brad had opened a can of corned beef for him.

We made our return trip to the Cat in about twenty minutes without incident, while the ROK's of the Tiger Division had to walk back to base.

Two Pounds of Flesh

I still had duty that night, but an hour or so before guard mount, Barry and Joe came around shouting that we needed to go over to the Korean Camp.

"What the hell are you talking about?" Vinny threw down his cards.

"You won't believe what is hanging from their main gate."

"Stop with the games, man," I said, "What are you talking about?"

I was getting more pissed by the second as Vinny folded his cards and jumped up, leaving me the only one at the table with the best poker hand since I had learned the damn game.

Another young man entered behind Barry and Joe. "Check out the main gate at the ROK camp. They hung two bodies from the entrance." He ran out but shouted back to us. "The Base Commander is upset and has requested they remove the bodies immediately."

We pulled our pants and boots on and took off for the camp, only a short distance away. Sure enough, up there over our heads for the whole world to see were two bloodied bodies we couldn't tell were even human. They were VC killed by the very same soldiers I was in the field with just a couple days ago.

"Joe, what's the story on this. Do you know anything?" I asked.

Joe carried his camera around all the time now and had taken some pictures of the mess hanging from the gate. The thousands if not millions of flies attacking the raw meat of the flayed VC were nasty in their delight.

"Base Commander is madder than a hornet's nest. He told the Korean Camp Commander to remove the bodies."

Joe was so excited he dripped in sweat and stuttered, trying to smoke at the same time. "You won't believe what the ROK Commander said." He just stood there, not telling us.

Vinny and I shouted in the same voice. "What did he say, Joe?"

"He said, 'No fucking way; they can rot where they hang.'"

"Fuck you, man," I said. "No shit?"

"Swear to God I'm t-telling the truth," Joe said.

"So where did the bodies come from?" said Vinny.

"I don't have a clue. All the talk is about the words exchanged between the Commanders. It was hot and heavy, but the Korean Commander was standing his ground with our Base Commander, not willing to give an inch."

"Well," Vinny said. "As cool as all that is, I'm tired and headed for some sleep. This is just too much excitement for me."

"Goddamn Vietnam. Do you believe this?" I said. "This gook-on-gook torture is some gruesome shit."

"For sure we will get some more information at guard mount," said Joe.

About thirty of us were standing around waiting for guard mount to start, dressed in camos or utility fatigues that were dirty and some downright smelly. We were in full battle dress, meaning we had our weapons, helmets, and web gear. Radios were turned off. The only thing missing was the dogs but their presence was heard.

Someone yelled. "Fall in."

We never really fell in. No one gave a shit.

Sergeant Francis shouted. "Attention." We came to attention. "At ease, men. I know you can't wait to hear the latest news on the bodies still hanging from the main gate at the ROK Camp."

We strained to hear every word over the barking dogs.

"We had a request two days ago to provide a team to go out on a re-con with the ROK Army. Airman Moore went as backup for Brad and his dog Silver. There was one serious alert that resulted in a small amount of ammo and rice that was still warm but the VC had left. It took about five hours to clear all the structures in the area, and then our guys were loaded up for the trip back to base."

"Airman Moore, is that about the way it happened?" said Master Sergeant Johnson.

"Yes, Sergeant Francis nailed it. We proved what our Sentry Dog teams can do."

"That is where our part of this story ended until this morning," said Francis. "I gather most of you have seen the two bodies hanging from the main gate at the Korean camp?"

He held up a hand for silence. We were on the edge of our underwear in anticipation and the dogs were all ears, too.

"While our men got a ride in on a Pedro, the ROK soldiers started walking back to their base. At nightfall they were ambushed. They managed to kill one VC and had several blood trails exiting the ambush site. In total darkness, they searched but didn't find their own missing soldier or the wounded VCs. In the morning, they had their two dead and three wounded airlifted out and went looking for their missing buddy. By evening they caught up with the VC, only to find their comrade, where the Cong had stripped the young man raw of his skin and left him tied to a tree. He had been reduced to a bloody pulp that was covered with flies."

The Airmen were buzzing. I felt a slight stirring in my belly at the thought of what my fate could have been.

Now it makes a lot of sense why two dead VC are hanging from Thunderbolt, the ROK Troopers main gate.

"All right, hold it down," the Sergeant said. "During the firefight the ROKs managed to capture two of the VC while killing several more of the enemy. A couple ROKs were wounded and air lifted out.

You can see the handiwork of the ROK Army hanging over the main gate. This is some nasty business, and the ROK commander is still refusing to remove the bodies."

The bodies hung out in the hot sun for three days before the Tiger Commander relented and had them removed. Like all VC killed in action, the bodies were taken to the little village of Phu Cat or downtown Qui Nhon and dumped in the street to send a message to the people.

That's two less enemy coming through the wire and trying to cut our throats. Fuck 'em.

Brad and I told and retold the story of our adventure with our fellow handlers. Being out in the field with those ROK Troopers and facing the same dangers that could break out at any minute reminded me I wasn't prepared or trained for real combat, the kind that turns men mean.

I agreed the death of the enemy, anyway it came, was good. It would not leave my mind any time soon, the vision of two bloody lumps that were unrecognizable as bodies, covered with flies, and hanging from the main gate.

Real Dog Initiation

A week later, after guard mount, I was viciously attacked by one of our own K9s.

I started towards Junior's kennel. I was attacked by one of the really bad-ass dogs. We are not to tease dogs, period. I kicked at the kennel door as I walked pass one mean mutt, expecting to hear the growls and barks coming from a very upset dog as he hit the (unsecured) chain link gate. I didn't expect to have my pants ripped off my body one second and the next, a jaw locked down on my thigh with four K9 teeth firmly attached to a bleeding thigh. I'm on the ground spinning in a circle trying to detached the dog from my leg.

I yelled to the handler, "Get this damn dog off me."

The Airman had to choke-out his dog to get him to release.

I bled a lot, got a night off, and the wound made a nice picture. I never messed with another man's K9.

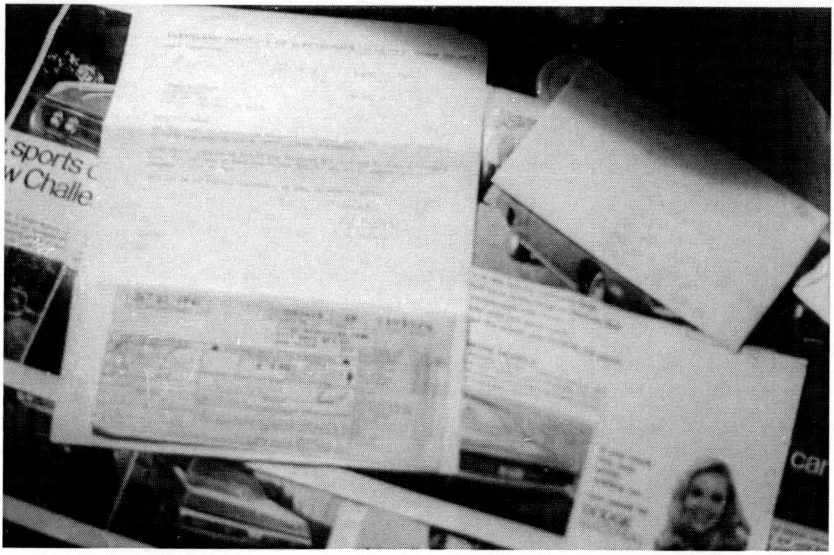

Chapter Ten

Back to the World

Phu Cat Air Base, Vietnam – November 1969

US Headline News

John Lennon Returns his MBE Medal to Protest British Government's Support of US War in Vietnam.

Sesame Street Premieres on Public Broadcasting System.

In Vietnam the war grinds on. No let up to all the dying. Back home it's football season, Thanksgiving just a month away. I'll spend my first and last Christmas at the Cat. In two months I'll return to Cam Ranh Bay and catch my very own freedom bird for the world and life. In the trenches, so to speak, life had slowed to a crawl. The assumed threat was always with us, always. No enemy attacks in the last 100 days were told at guard mount one night. Really, we didn't know. As the earth tilted northward, the southland got even hotter. There were nights of hundred-plus degrees and you'd beg for rain.

And when the rain finally fell, I knew I had never seen such weather as this in Florida. When the rain stops then teeth shattering cold rakes your numbed body that exhausts the mind. We always showed up and did our duty. The only thing giving strength was going home. I officially was a SHORT-TIMER. Soon I would lift Junior to the back of a deuce-and-a-half for our final ride. I didn't feel leaving him was a big deal. But my heart told me different.

Side Note - the story has slowed. The NVA, the VC, the gook farmer, whatever, stopped coming to our playground. The chatter among handlers said fewer alerts. Over one hundred and twenty plus days without a peep! Even the snakes had left dog and man alone. Glory. The question was, 'What's Up Doc?' The big shots were nervous, so the workers had to be nervous. We, who put our lives on the line every night, were happy! So, the brass sends lower brass to remind us to be alert, be nervous men, our friends are saving up for a big hit. Continue the excellent job you do in the night, you handlers do it so very well. Then he was gone. Staff Sergeant Francis had shipped out a few weeks ago and the new guy felt it necessary to repeat the same message. And it worked, we all became nervous.

On post a few nights later, Joe came calling, bringing his own bucket along. He was nervous also. So that night we enjoyed fresh popped popcorn, soda that was sort of cold, and TV.

"Joe, what else is in your bucket?" I wondered what the bunker guards thought hearing snap-crackle-pop out in front of their post. I know their mouths watered at the smell of hot oil bursting open corn. We put away the small TV, too distracting, and the dogs became too relaxed.

Coffee run was early and after, Joe returned to his own post. Later, he told me he had only thirty minutes of battery playing time left and the weight of the thing made it a one-time experiment. Popcorn was great for anxiety relief. A good movie would have made the night perfect. The moon was bright, the night sky was clear and the sweat and mosquitoes ever present. Oh, FYI time! A cure for mosquitoes is at least a seven mile per hour wind. A cure for the sweating was to remove your uniform top and let the new winds air-condition your bare chest and back.

The minute the winds stopped, you quickly buttoned up. The winds were a blessing. The mosquitoes were nastier than any bug in Florida.

By now, the calendar said I had just forty days left in this man's war. God! I want to go home and hug my mother and father. It saddened me knowing there would be no girl to hold. I would think serving almost a year in Vietnam would have been long enough to quell heartache.

I had thoughts of staying at the Cat and serve the rest of my enlistment. Yes, like Rev, but my next duty assignment was in stone, I'll be going to McGuire AFB in New Jersey. No dogs.

Friendly Fire

One night it was raining kittens and puppies as Junior and I dismounted the truck and headed for our post. This post and the one next to it were heavy with vegetation. My favorite, the same post I covered my first night walking Duke II. Tonight, I didn't really know the handler next door, so I would spend the night alone.

The cool thing about this post is the Buddhist temple, about eight feet by eight feet, hidden there in the little jungle. I'll clear my post and make a beeline for the temple. There's enough room for two teams to get in out of the rain, but I expect I'll be the only occupant tonight.

First, I had to clear the temple using my flashlight. I lit up the interior, checking for anything that would bite me, like snakes, spiders, scorpions, rats, and the list goes on. It passed the first test; the second test was letting Junior make his inspection, and upon his approval, we entered and settled in for the night.

I was chilly from the rain and kept my poncho on until I heated up some and had to take it off. The damn things didn't breathe at all. The rain poured from the sky and lightning lit the night up, making for a show of nature's powers, and I had the best seat in town. It poured down hard until after the coffee truck had come and gone. It was late and the first shift and the relief truck where on their way.

I don't like coffee, and making small talk with the driver will just delay him getting the hell out of here and back to the dry comfort of the kennels.

After a while, the stars came out, then the moon, and the area was lit up well enough to see some of my post. I stretched my weary legs and enjoyed my shelter from the rain. A strong breeze blowing into my face made me feel more secure, because wind in the face was best for our dogs' noses. In these type of conditions, Junior could pick up scents from miles away.

The rain stopped, so Junior and I cleared our post again. I found some water that I later discovered was part of a creek, so I retrieved my bucket from the temple and sat down by the stream. From my c-rations, I opened a can of peaches and a can that contained pound cake, and what a combo! Together it was heaven.

All is quiet, just as I like it. No planes in the air because of the weather, except for Spooky, who's in the sky most every night and all night.

I heard a shot and a splash in the water about four feet from me, and I about shit in my pants. I didn't move or do anything. I sat, waiting to see what happened next. Nothing.

I thought of the story the Rev had told me about friendly fire happening to him one night. I knew there were ARVN soldiers, our Vietnamese allies, guarding the bridge that crossed over the same creek, so I waved and waited for the sound of laughter to drift down to me. When it did, I was relieved.

They know K-9 is on duty. I can imagine looking through a Starlight Scope and seeing some dude moving around with his dog in their area of responsibility to guard. Hell, it's probably fun to scope us out. I hoped it was a challenge. I wave again with thumbs up. I heard laughter. Is this how legends are created?

Thanksgiving, 1969

"What's your name and where you from, I mean, back in the world?" Vinny said. He extended his hand to an Army man who had come to Phu Cat to celebrate Thanksgiving with the Air Force. Isn't that special!

"Benjamin, Benny Weaver. I'm from Texas, actually, west Texas. My unit just came in from the bush to rest for a few days. I'm a Scout Dog Handler."

We could tell this brown-eyed grunt was beat; his deeply tanned face was streaked with dirt and sweat. His uniform was filthy and so was Benny. He needed a haircut, but we were not going to tell him so. We all introduced ourselves and asked if he was hungry.

"We're going to stay long enough to enjoy that famous Thanksgiving dinner you Air Force guys serve up. We can't wait. I'm so tired of eating c-rations I could vomit."

We adopted him right there, and our first order of business was to get some Air Force food in him. We all headed for the bus stop, boarded the blue bus, and continued getting to know this Army Dog Man better.

"Ben, how long you going to be with us?" said Barry. "Do you need a place to rest up while you're here?"

Joe chimed in. "We happen to have an empty bed in our hooch, so you don't have to worry about where you'll be sleeping."

We entered the Airmen's Mess Hall and got in the short line as the cooks served up whatever we wanted.

"Benny, this chow hall isn't as fancy as the NCO open mess is, and we would rather eat there all the time, the food is that much better," said Barry. When it came to food, he knew what he was talking about. Barry even knew what food specials were on what day.

Benny asked for a little bit of everything. He piled on the bacon and eggs over-easy. The chipped gravy over biscuits went on a separate plate. He grabbed a box of Rice Krispies and another tray, then a glass of milk and a couple of sweet rolls.

Benny distributed the food between the two trays, then carried one on each hand back to our table.

"It's been months since I've eaten like this," Benny said. "When we found out we were coming here, we got so excited, we almost got ambushed. If my dog Jacket had not alerted and warned us, some of us might have died. That was two days ago. It took us this long to catch a ride here."

Vinny spoke up. "Hey, you guys leave the man alone and let him enjoy his food, all right?"

We had a new respect for the Airmen's Mess Hall and the food tasted better than it had ever tasted. Watching Benny eat grounded us.

We have it better than most people in Nam and for sure a whole lot better than this dog grunt.

We finished our plates, except for Barry, and the rest put our trays away and returned to the table. When Benny finished his first helping, he returned for an order of pancakes. Piles of bacon and sausage patties looked like a small hill on his plate.

We've been taking this all for granted.

I looked around for Vinny and found him talking to one of the cooks. He returned to the table.

"Hey, Benny, see what you think of this," Vinny said. "The Mess Sergeant said he could find a steak for Jacket and prepare it any way you would like, or if that doesn't suit you or Jacket, how about some bacon and sausage? Whatever you want."

"Man, that is a great idea. He hasn't had a real treat like this in a long time. He will go nuts."

Ben wanted to take some bacon and sausage with him now and the mess cook was glad to make it happen. With a large portion of Air Force food in him, he was looking better and a lot more talkative.

"If one of you will return with me to the kennels, I'd like to drop it by for him now. Thanks, man, this means a lot to me."

I went back to the kennels with Benny and watched his dog Jacket devour a pound of bacon and sausage. Benny took Jacket to the exercise yard and let the dog run and play at his leisure. They carried on for a half hour or so before Benny was ready to put his dog up and get some shuteye.

It was almost 1200 hours as we entered the hooch, so everybody was racked out and snoring. Someone had gotten sheets for Ben's bed and made it up. The look on his face was the same as his look in the chow hall. Wonder.

"I'm so used to sleeping on the ground, I'm not sure if I'll be comfortable in a bed with a real mattress."

Reminds me of why I joined the Air Force. Only us Airmen pukes get to sleep in a bed with sheets, pillows, and a blanket in Vietnam. Shame on us taking it all for granted. I wonder what his reaction will be to a hot shower?

I had early shift that night, so I didn't get a lot of time to sleep before the pickup trucks arrived for the trip to the kennels. I noticed Benny was sound asleep.

So, Ben's adapted to a mattress easy enough. Sleep on buddy, see you in the morning. Maybe we'll play some cards and get drunk.

At guard mount, Sergeant Mendez, the new guy, gave the Nightly Action Report for all SEA. "Several air bases got hit during the night, but no Air Force KIA's or WIA's. Six NVA were KIA and two were captured. Da Nang took most of the action, with four enemy killed and several aircraft damaged. Two more NVA were captured there."

The wars in Thailand, Cambodia and Laos were long, nasty and deadly. What war? DUH!

Who do you think are American Military Service Men and Service Women?

Most don't care about uniforms anymore. Hada mighty knows, isn't that special. It was party-time in the so-called free world. But SEA had to be tamed. It took brave men to put on the boots and walk the job. I know what Smedley Butler says, he's a United State Marine, enough said. The real drama is in the lower ranks of enlisted men, mostly the poor hoping to gain an edge through education. A Vietnam Veteran longed for a better life through knowledge. To be smart (intelligent) enough to help make change means something. People, friends, listen up, America: the real hero stands for a truth and the principles behind it. One simple word. Freedom. We are the only beacon that can make a difference. The Vietnamese think we are all rich. They think, 'Why come this far to fight in our war, to fight our battles?' Think about it? Are Americans perceived as different from the rest of the world? Oh yeah, big time. American military personnel are the reason and the saving grace for the People, our people. As a nation, we over produce BS in a big way. When this country goes to war, it's done BIG. When you have poor people seeing American blue-collar young people in uniforms and here to secure freedom for them, are you kidding me? Many say go home, American not needed. The blue-collar Airmen, Sailor, Soldier, Marine and Coasties are the force behind the backbone of our military, our nation. And our gentler side, believe it or not. They get the job done in uniform and out of uniform. The Young Men and Women in the uniforms of our country are the best ambassadors this country has. And, we the people, reap the dividends. They are your neighbors, they look just like you, Millennium's. Most service personnel do not face danger getting that wanted education, the poor have to face blue-collar task. A task that only they will do, if ordered and that can kill you. All this for wanting a better life. Hell yeah!! That's a Vietnam Veteran. Mostly I'd like to think America sent a hippie Army to SEA. Most of us never fired a shot. We wanted to start a garage-band. I knew at least half a dozen handlers who drummed, Vinny being one of them. We took drumsticks hidden in our buckets to pound on the hard clay ground. I'm telling you, America sent its very best hippie army to SEA. Be GLAD you did! Our young people are happy to be alive, to solve problems and grow-up sane, productive humans. That is a Vietnam Veteran. The spirit of freedom in our veins is the same. Young Airwomen and Airmen are good energy. Young people, Babyboomers gave the internet to the world and made knowledge available of freedom's true intent, bringing freedom to the world. There is so much to change and it must be done now! Millenniums come together with Babyboomers. Let Boomer's finish what they started, let's give The New Millenniums a big shove. A good name for a country band.

Vietnam Veterans know how to face-off with life's issues. Well, sort-of, anyway. Most Vietnam Veterans have it together but other's hurt themselves (22/day).

◊ ◊ ◊

December 1969

US Headline News

John Steinbeck, American Writer and Nobel Prize Laureate, Dies.

First Draft Lottery in US Since WWII.

Altamont Free Concert Held at Altamont Speedway in northern California.

Hosted by the Rolling Stones, it is an attempt at a "Woodstock West" and is best known for the uproar of violence that occurred. Altamont is viewed by many as the end of the sixties and the summer of love.

◊ ◊ ◊

Early January 1970

Farewell, Good and Faithful Friend

Our little group of nine had been short-timers since October, and were counting down the days until at last, my next to last night on duty in the Nam. Earlier, we had packed our gear in wooden crates for shipment home.

Is this actually going to happen? I have my doubts.

We talked a line of shit a mile long, we rubbed it in. "We are going home, thank God!"

"We know, we know, you're going home," the new FNGs would say. "Good fuckin' riddance."

Christmas and New Year's was nothing compared to going home. Our Army Scout Dog Handler friend stuffed himself on Thanksgiving, all day long and made several trips to the kennels with food. He was already gone the next morning before the first shift rolled in.

I didn't consider my feelings about leaving Junior behind until my last night of duty. Junior sat on his ass and slumped down at the shoulders. The sadness in those eyes woke me to the fact that I was going home but Junior wasn't going anywhere. He knew I was leaving him.

"Damn, Junior, I'm going home buddy. Can you dig it?"

Junior will die over here and no one will know the difference or care. The folks back home will never know the kind of work he performed night-in, night-out. None of these dogs will return to the U.S. What a fucking waste.

The only death on my hands is a good dog that wouldn't play by the rules of war. Too bad something better couldn't have been done for him. But that's Vietnam and that's the Air Force way, screwed up.

I scratched him behind and in the ear, like he loved. "I'm so sorry, Junior, I hate leaving you behind."

Junior lay his head on my knee, so I bent down and wrapped my arms around him, held him close to me.

"I had forgotten that you have feelings, too. Forgive me, boy. If the Air Force would allow it, I would take you back to the world where we could continue to live."

Don't kid yourself, Jimmie. You won't miss Junior any more than you'll miss the Cat or any part of Vietnam. You hate everything about the place, the countryside, the rice paddies, the mountains, the deformed-looking people.

All of us handlers and dogs are like engines on a plane; we run and run until we give out. And like an engine, if we catch a bullet, we're replaced and written off as the cost of doing business...I mean war.

Before we were dropped at our posts that last night, I made eye contact with Vinny. He shook his head and lifted his weapon in a sort of salute. I knew what he meant.

Is this the night we get hit again? What if we end up in Hotel Hanoi? Or get killed with one working night left in the Nam? Or the absolute worst, our orders are canceled because the needs of the Air Force come first. A group of nine handlers leaving at one time puts a big hole in the duty roster.

All in all, it was ideal conditions for Sir Charles to visit. Only stars would light his path.

I'll probably drive poor Junior crazy tonight and overwork both of us looking for an enemy who just might be out there, but probably not. Maybe they heard about a group of nine handlers scheduled to ship out and want to give us a going away party. Who knows? Glory!

Word or two about our Noble Canines

For most Handlers, it was an honor to have worked a Sentry Dog and to have fallen in love with the breed. German Shepherds are at their best, protecting. My training dog Duke I at Lackland AFB, had trained three different new guys. That's twenty-four weeks of obedience training. Everyday, except weekends for the first four weeks, we trained eight hours a day on nothing but commands and they did them and taught us. Be impressed. Remember the numbers next time you send your dog to another Obedience School. If you don't do your part and continue the instruction, the lesson will melt away and you will be crying for a refund. Remember, what goes down the leash, comes up the leash. Or as Forest would say, 'Stupid is as Stupid does.'

I was leaving behind A Mighty Canine, Smoky-X480 (Junior, to his friends). I was proud of my Noble Hero. Once we got past the humping part, we settled in like a couple. He loved c-rations. He loved to have his belly scratched, behind the ear scratched and a little in the ear was better than rump scratching and flank scratching combined. He would lay on a side and lick my hands, as I toweled him dry. Sometimes, by the time I finished brushing him, Junior would be out like a light. No card game for the boys and girls at KatKinnelKlub, NO BOOZE served.

I left a healthy Junior on January 5, 1970 and by the last days of December 1970, Smoky-X480 was no longer a GI. Disease took him. I never stopped to considered Junior was as exhausted as I was.

In jungle warfare, MWD's are a good fit. Desert warfare, not so much. Dogs and handlers who are putting life and limb on the line detecting bombs buried in roads and pathways are a waste of a good drone. Dogs are good PR; Shepherds, Powerful and Noble.

◊ ◊ ◊

I waved to Vinny as the big deuce-and-a-half pulled away, leaving him standing on a clay road that ran the length of a ten-thousand-foot-long runway. Us, we, all handlers, were made to man-up one more time. Who else would understand the edge but Handler and Dog, sixty-two teams, every night. Who else but a nineteen-year old boy-man who was dumber than a box of rocks and full of cum would walk a barren wasteland for ten hours every night, who? I'll tell you, the Sentry Dog Handlers.

I've marked off a whole year of putting my life on the line for folks back home who don't give a rat's ass. Think about it, I got six nights off. Two of those nights off included the trip to put Duke II down.

Last star light at the Cat and time slowed turning seconds into minutes. Finally, the end of another long night. It was time to put Junior to bed.

Is this what it would have been like to say goodbye to Lillian on my way to war? Oh, how it works on me not having my special girl waiting for me.

"Thanks, Junior," I said, "for getting me out safely."

Back at the barracks, the card game was just getting started and I was ready to blow off steam. We had one last job to perform for ours friends and we would not forget, but for now it was time to drink ourselves silly. We had made it through our 363rd day.

"Hey, Moore, what in the world made you decide on New Jersey?" said Bolden, from the other end of the table. "I thought you would head back to Florida. What's up with that?"

"Yeah, Jimmie," said Vinny. "I still can't figure it. I'm from Jersey and no way do I want to see snow again unless I choose to."

"I want to be around a lot of people again," I said. "I want to go to different beaches, get drunk, get a serious tan, and hang with lots of girls. Maybe see New York or Philadelphia, maybe even Washington. I still have memories of a lost love to get over and fighting this Man's war hasn't made it easy to forget."

Joe was running around getting pictures of beer cans stacked to the ceiling, the black and white TV we donated, and our Christmas tree, as well as whiskey bottle ribbons attached to the light fixture, from the hundred or so bottles that had been consumed during our year in hell. Vinny owned up to stealing that old light fixture on his post one night when we first got in country. Light for our card games. That's how we caught Barry with an ace up his sleeve.

We had been playing cards and drinking whiskey and beer for hours and were still too excited to lie down for some sleep, so off we went to the Airmen's Mess for some lunch.

We made plans to pick up steaks from the mess hall early in the week, and the mess Sergeant had them ready in nine individual bags. Now all that was left was paperwork and goodbyes.

At the kennels, Junior could hardly contain himself, so I tossed him his steak to avoid him beating me to death with his tail.

"Look what I've got for you, good buddy," I said. I seated myself on the warm concrete. "Look here, boy, doesn't this smell good?"

"Damn right it does, it don't come any rawer."

I'd never seen Junior acting so crazy or so playful. Junior had the steak in his mouth and was strutting around the kennel like he was royalty.

"Come on over here and sit down." I slapped the floor to get his attention. Junior began devouring the steak in huge gulps.

"Slow down, Junior, you're going to hurt yourself."

In a few days I'll be eating coconut cake and hearing Mom say the same to me.

Junior was finishing off the bone that he held between his paws. I smiled, watching him enjoy himself.

"Well, boy, this is my last day at the Cat and my last day with you. Heck, I don't even get a chance to give you a good brushing, sorry buddy."

Junior lay down next to me. Resting his head on my leg, he looked up at me with big brown eyes.

"Junior, I'm going to miss you."

His eyes held sadness and I started to choke up. To distract myself from crying, I rough-housed him and got him to wag his tail.

Now I was weeping. Junior knew it was goodbye time.

If the guys see me now, there'll be no end to the ragging. Hell, I bet they're boohooing like babies, too.

"See ya, Junior. Thank you for getting me through this year, my war."

The last sound I heard as I walked away was one final whimper from my best friend.

A First at the Cat

I met Joe at the bus stop. "Hey, man where is everybody?"

"Gone, man, where do you think? Everybody's in a big hurry to get on the freedom bird, I guess."

"Joe, do I see crying eyes on my friend?"

"Go fuck yourself."

"Come on, Joe, it's okay. I cried, too. I'm going to miss that dog. We spent a year with these animals."

"I know, who wouldn't cry? I'm a softy," Joe said. He struggled to keep his lower lip from quivering.

"Joe, this may be the last time I ever see you. Isn't that a load of shit?"

"Ah, I hear ya," Joe mumbled. "It's been a good year in some ways and getting to know you was part of that."

"Me too, buddy." I said. "You and Vinny are going to freeze your rears off back home."

"I guess so, but it will be good to get home and do something different."

"All I want to do is sleep and eat some good-ole home cooking. Beef stew comes to mind."

"Beef stew," Joe said. He turned to face me. "Never had it before."

"You come and visit sometime. Hey, come with Vinny. You know he is going to try and make it down and we will drive back together. That would be a hoot."

"I've had enough of warm climates for a while," said Joe. "In the next thirty days I will soak up New York snow and ice before I check in at Eglin."

"Isn't this some strange shit, me going to New Jersey and you to Florida?"

"Really."

Everybody in our Red Horse hooch-housing was going to be flying out tomorrow. I lay down and thought how strange it was to have nothing to do and no place to go. I closed my eyes to let my mind run where it would go. Before my eyes appeared a big jet, and it was on fire.

"Shit!" I jumped up. "Anybody want to go to the club?"

"Vinny is already there," Joe said.

"Are you going?" I was already putting my boots on.

"Catch up with you later."

At the Airmen's Club, Vinny, Teagan, Puls, both Mikes, Bolden and Malone were drinking beer, and by the looks of the table they were well on their way to an all-out drunk.

"I just stopped by to see if anyone is interested in some food. I'm starving and I don't want to get drunk on an empty stomach," I said.

"Sounds like a good idea to me," said Vinny. "I'll get you a beer."

He got up, headed for the bar, and returned with a beer for me. "We've got a long night of drinking to do, and it's time to get caught up during normal drinking hours."

I was glad for the food break because we did some serious drinking.

I wanted to wave bye to the troops as they mounted up and headed out for another night into the Twilight Zone, but I didn't make it because after the tenth beer I didn't care and neither did any of the others who were returning to the world.

They'll just have to make it without the fearless nine.

I went to sleep drunk my last night at the Cat, dreaming of the big bird waiting for us at Cam Ranh Bay to take us back to the world and home and safety.

Much later in the night someone was shaking me, trying to get me awake.

"Jimmie! Jimmie, man, get the fuck up, we're getting hit," Joe said.

"To hell you say? Leave me alone, I'm trying to sleep."

"Get the fuck up, we are getting hit, shithead," said Vinny. He was on his way out the door.

"Fuck it, I'm getting out of here," Joe said.

At last there was silence.

Did somebody say we are getting hit? Fuck it. I'm done with this war.

Within a few minutes, I was fast sleep.

Again, hours later someone was pounding on my bed, yelling at me to get up and get to the bunkers because the base was under attack.

"I know, I know. Leave me alone."

It was about 0930 the next time I was awakened. I was glad to see sunlight because I was going home today. I had made it through my last night in the Nam at Phu Cat Air Base and I was going home.

"Going home today," someone yelled.

All eight of my crew were returning from their night in the bunker. Joe carried the mattress he'd brought from his bunk.

"You dumb-shit, you could have gotten yourself killed last night," said Vinny.

"What the fuck are you talking about?"

"We got hit by NVA rockets just after midnight and again at 0600."

"Yeah," said Joe. "That was the f-f-first time the Cat's been attacked by rockets."

Vinny's brow furrowed with aggravation. I wiped the grin off my face and inquired if anybody wanted to go to breakfast with me.

Joe threw a pillow at me as he replaced his mattress back on his bunk.

"What is that all about, you taking the mattress with you?" I asked.

"I'll tell you over breakfast, let's go. Last meal at the Cat."

"Got that right."

On the bus to the chow hall, Vinny said the attack began sometime after 0100 hours, when the NVA, camped up in the mountains, started dropping rockets onto the base, including flight line and housing alike. Within minutes, everyone on base was heading for the bunkers.

"We tried to get you up, but you drunken dummy, you were passed out cold. We had to leave you," Vinny said.

"That was nice of you guys," I said. "Well, what happened?"

"Most of us got on the trucks and headed for the kennels or a bunker," Joe said.

"Yeah," said Vinny. "By the time we reached the kennels, I decided to get my dog and gun and spend the rest of the night riding in the back of the truck as back-up."

"No shit, you got your weapon and dog and rode around for the rest of night?" I asked in amazement. "What about you, Joe? Did you get your dog and go to work?"

"No, I didn't," said Joe. "I started to, but decided I had enough of this flaky-ass war, so I got a ride back to Red Horse and moved me and a mattress to a bunker to sleep for the rest of the night. I had no idea we had been hit a second time until Vinny woke me."

"They hit us again with rockets around 0600 hours. I have no idea what kind of damage was done to the base or if anybody got killed," said Vinny. "There was no ground attack, just rockets. They sounded like an eerie buzz, then it would stop, silent, before an explosion."

"How many rockets hit the base?" I asked.

"Not many, maybe four or five and then it was quiet." Teagan spoke up from behind me.

"How close to Red Horse did they come?" I asked

"No closer than a couple miles. You were completely safe in your bed." Vinny said. "So were you Joe, sound asleep on a mattress in your bunker, you're nuts."

As we consumed a large breakfast, my friends laid out all the facts. One person on the flight line had injuries, but they were not serious. The rockets set a few buildings on fire and the flames were extinguished. One aircraft took minor damage, and that was it.

I missed it all and so did Joe.

Not too smart of me. I could have been killed where I slept. Fuck Vietnam.

By the time we got back to Red Horse after breakfast the rest of the handlers were returning from their long night of duty. We understood when most hit the bunks without a single "goodbye" or "go fuck yourself."

Overjoyed and acting silly, we said a few farewells but we let most of them alone to get some rest. They had to do it all over again in about ten hours.

In our 365 days in country, we lost one handler who was wounded and sent home, and my dog, hippie dog Duke II to non-workability, death by injection. We had been hit three times our entire tour of duty, but the Cat was now on notice that the commies had the means to deliver rockets. Reach out so to speak.

From the Cat to the Bay

I had little in the way of luggage because we had already shipped most of our stuff state-side. I carried a small gym bag with all my remaining possessions and wore my Class A uniform for the ride home.

The Loadmaster began to close the giant doors to the C-140 and take-off was just moments away. We held our breath as there was always the chance of a rocket taking the fragile bird out of the sky. Downing an Air Force transport plane full of military men preparing to return to the States would be a big win for the enemy, so we all sat a little more puckered in intense anticipation.

The hot as hell aluminum coffin-plane began to crawl down the flight line toward the end of the runway. We would lift off into the wind going east and I would get a quick view of the landscape as we picked up speed and height.

"Hey guys, there will be a most interesting view of our kilo post as we move down the runway." I yelled over the roar of plane's engines. "Check it out."

We peered, transfixed, at the bleak ground we had patrolled for the last year. It looked worse than we'd imagined and hit home just how exposed we were, easy targets for a sniper or sapper.

As our uneventful lift-off continued, I knew if I strained looking straight down the side of the plane I might get to take in Kilos 54 and 53 at the end of the runway. Kilo 54 was my first duty post. I didn't see much because it was lush, green jungle and the C-140 was gaining altitude.

"Look, Joe, can you see the ROK watering hole?"

All this time and we had never seen this small camp that the ROK Tigers had used to pump water from the ground and transport to their main camp. I wondered why the Koreans just didn't use our water that was treated and safe to drink.

Within fifteen minutes we were over the ocean and boy, was it beautiful. Safer, too, but we would face the same peril taking off in the morning.

At Cam Ranh Bay Air Base, Vinny, Joe, and I found bunks for the night in a blistering hot, half-tent, half-wood frame building. We had hoped for air-conditioning. There was a lifer putting his gear away and he didn't even acknowledge us as we came in. We stored our gear for the night and started out the door going to the mess hall for supper.

"You boys better stay away from my property." The NCO scowled. "I'll be taking my money with me and my camera, so there is no reason for you to be in my gear. Are we clear here?"

Joe spoke up. "No problem, Sergeant, we're not thieves. We just want to go home in the morning."

Vinny and I were angry. "Fuck this son of a bitch," said Vinny. "Let's get out of here."

"What did you say, Airman?"

"I'm out of here…get me away from this son of a bitch."

"Airman, get back here." The pissed off Master Sergeant was screaming. "Get back here right now, or I'll call Security Police."

There were six or seven of us who yelled at the top of our lungs. "We are Security Police."

"Eat shit and die, you lifer bastard," said Barry.

This incident was not going to put a damper on our getting shit-faced, but we had to hold Vinny back from going and beating the shit out of the lifer. By the time we got to the Airmen's Club, we made a promise to leave it at the door and just have some fun our last night in Nam.

The place was almost empty and the bartender kept giving us free beer. I think we paid for every other one or something.

Several of us left the party early and returned to our tent ready to call it a night. The lifer was sound asleep and snoring.

This is a good thing, I hope. Maybe the Sergeant will let bygones be bygones.

I climbed into my bunk and lay thinking about Vietnam, Junior, going home, Lillian, my friends at Phu Cat, my next duty assignment at McGuire, Vinny and Joe leaving for bases in Florida and California. I started to feel sad and lonely, but then I heard the rest of my friends returning from the club.

They were singing, "I want to go home," loud enough to wake the sleeping lifer.

I jumped up from my bunk and ran out of the tent, hoping to calm them down before they woke up the Sergeant.

I used my arms to motion for quiet, but they became louder and laughed their asses off at me for even trying to shush them because they were all sloppy drunk.

"Is the lifer shit sleeping?" Vince said. He was slurring his words.

"Yes, he is and let's keep it that way," I said. "Be quiet and let the man sleep. It's late, so get to bed. We have a plane to catch tomorrow, and none of us want to be left behind in a Security Police holding tank, for God sake, so shut up and get to bed."

By the time we entered our tent, calm had returned, or at least it seemed so, but I knew Vinny would have no problem awakening this guy and starting where we left off earlier. I pulled Vinny to the side.

"Vinny, the lifer is asleep." I whispered in his ear. "Maybe he will let our verbal attack go away with a good night's sleep, so let's not fuck it up, okay, buddy?"

The next thing I knew, Vinny was standing over the lifer holding a red fire extinguisher in his hands high over his head. He was ready to have it come crashing down on the man's skull.

"Vinny, NO, don't do it."

I leapt to his side and pulled the deadly device from his hands while pushing him back to his own bunk. Then I started to take his boots off. Vinny still hadn't said a word.

"Vinny, you need to cool off or this lifer is going to have your ass in jail. I've seen LBJ, it will ruin your life and no telling when you will get home."

"Vinny, listen, listen buddy." Now his eyes were closed, so I shook him. "Do you want your pants removed?'

He made eye contact with me. "Oh yeah, and will you blow me, too?"

I heard a snicker from behind me. It was Joe giving me all his wasted moral support for this little dilemma.

"Come on and get to bed. We've a plane to catch and sleeping will help pass the time." I stood for a while just in case he started to get up, then bent to hide the extinguisher under my bed. I was soaking wet with sweat and needed some relief, like a beer, to cool me off, but the bar was closed. Babysitting Vinny was all that was on my mind.

"Jimmie, go on, lie down. I will watch him for a while."

"Thanks, Joe."

When I woke up, everybody was still sleeping except the lifer, who had left already.

I stared at the ceiling of the tent while the other guys slept. *I wonder if Vinny would have hit the lifer with the fire extinguisher. He was way too drunk for a twenty-year-old.*

Joe motioned at me from across the tent. "Jimmie, you think the lifer will be back with SP's?"

"Who knows, think we should put a watch out for cops?"

"Probably, but I'm going back to sleep."

"Thanks a lot, Joe."

At 0900 hours we were in the enlisted men's mess hall eating our last breakfast in the Nam. Our bird was scheduled to take off at 1330 hours. We had time to kill, but nobody wanted to go sightseeing around Cam Ranh Bay. The fear of missing our flight had taken hold and we were not budging.

Most of the planes coming and going were military, but we were all thankful that the plane we would be flying on was a 707 with pretty stewardesses, air-conditioning, and beer. Lots of beer.

I'll bet we're all worried about being shot down during takeoff. We've heard the stories of commercial planes shot down and men killed trying to escape this place. I'll distract myself thinking about how beautiful the sea looks and wondering what the stewardesses would look like without clothes.

By 1300 hours we should have been on board, seated, buckled up, and ready to go. Instead, we were still in the waiting area, waiting. People were getting anxious, then angry and I had the thought of a riot breaking out when they announced our flight would be delayed thirty minutes.

"Hey Joe, listen man." I was whispering. "How many people can they get on one of these jets?"

"I have no fucking idea," said Joe.

Vinny turned and said, "A couple hundred I guess, why?"

"Shit guys, there must be three hundred people standing around waiting to get on this bird. What do you think?"

"Damn it, Moore, stop bringing up this shit." This came from Barry, who never got excited about anything but was now waving his arms at me. "We are going home today. Charlie is not going to change a thing. They have seats for all of us. So, please, stop the gloom and doom, alright."

We stopped and stared at Barry as if he were insane. We had almost never heard Barry cuss or get excited about anything. He said "shit" and "damn," a sure sign of nervousness.

"Attention. May I have your attention, please?" The voice blared over loudspeakers. "We are sorry for the delay but we are now ready to begin boarding Flight 457, flight service to Seattle, Washington with a short layover in Tokyo, Japan. We will begin loading in ten minutes."

We K-9 handlers wasted no time getting in line, but to our dismay, the voice squawked again. "The order of boarding will be as follows: Seniors Officers, followed by Junior Officers, Senior NCOs, followed by NCOs, followed by E-3's."

We were being bumped as higher-ranking men began to move in front of us. The more we were bumped, always to the rear, the more nervous we became because at least fifty percent or more of the passengers were enlisted men with the rank of E1, E2 and E3.

Please God, no; you've got to get me, all of us out of here, please God.

"Jimmie, you got to cool it man," said Barry. "I promise you we will all get on board this plane and leave today. Okay, man?"

How obvious do I look? "I know, I know, just…can't help it."

Our move to the rear stopped, and we were about half way down the long line of men. We could all relax, knowing we had a seat come hell or high water.

Most of the ranking officers and NCOs choose to sit in the front of the plane. We moved to the rear and found seats so that all of us were together and within easy talking range.

It didn't take Joe long to strike up a conversation with one of stewardesses. He had Vinny take pictures of him with just about every one of the six stewardesses on the plane. They were willing because they knew these men waiting for the plane to takeoff had not seen an American woman in a year, some longer.

Joe took advantage of the opportunity to get a picture of a stewardess bent over right next to him, and from behind he got the picture looking up her dress. Vinny got the picture of Joe taking the picture.

We waited and waited for what seemed forever until the pilot came on the speakers.

"My flight crew and I would like to welcome you all aboard. My name is Captain Marshall, and I would like to put your minds at ease for the upcoming takeoff. I have never lost a plane taking off or flying the long distances that we fly to get this special cargo home. The 707 is as sound a plane as there is. The engines are starting up and we will be leaving in about ten to fifteen minutes."

As the plane started to move forward, all eyes were focused on the windows.

One last look at Vietnam.

The powerful whirl of the engines filled the cabin with our adrenaline. Faster down the runway we went, pushed into our seats until we could go no further. The nose of the plane began to rise, higher and higher until the wheels had no choice but to follow and we were no longer connected to the earth.

We are safe and going home.

Men laughed like they hadn't laughed in a long time. The "fasten seatbelt" sign turned off, and I got up to shake hands with just about everybody. I returned to my seat exhausted, fell into a deep sleep, and when I woke up I was in Tokyo, Japan.

Home

From Phu Cat Air Base to Cam Ranh Bay to Tokyo to Seattle to Chicago to Jacksonville to Starke, I traveled a grand total of 8,089 miles. My computer calculated those numbers, but they seem small to me now.

Vinny visited me in Starke for the first part of my 30-day leave, and we spent a lot of driving time in my new 1962 VW Bug. We chased girls at the University of Florida in Gainesville, even got dates, but the night we went to pick them up we had to confess we were not students at Princeton University in New Jersey, but Airmen just home from Vietnam. As if on cue, they both turned and left, taking the flowers we brought, our tokens of apology.

Neither Vinny nor I needed to play tricks and lie to get a date.

Next, we drove my VW to New Jersey to pass the rest of our leave time. I froze with the heat on full blast because Vinny would roll the window down. He was driving and couldn't afford to fall asleep.

New Jersey was freezing and McGuire Air Force Base got colder. I felt discombobulated, out of place here, so much so that I went to the Pentagon in Washington, DC to try for new orders to return to Phu Cat. But the Air Force people and the Pentagon just laughed at me and told me I would finish at McGuire.

After a year-and-a-half back in the States dating several ladies, I was still moping around about Lillian, but the Airmen's Club at McGuire AFB saved my life.

The night I met Virginia she was sitting at the Airmen's Club with a group of her girlfriends. She was wearing a curve-hugging sky-blue sweater with a dark wool skirt and boots, both attractive and practical for the New Jersey fall weather. Her green eyes met mine, lingered, and she winked at me from behind her blonde bangs. I winked back.

It didn't take long before I approached her, "I'm Airman Jimmie Moore, stationed at McGuire. What brings you here?"

"I'm Virginia," she said. She motioned to the others at her table. "I'm visiting a friend. I'm an RN and I work in Philadelphia."

A nurse! I've always wanted to date a nurse, especially after hearing what they did in Vietnam.

"Would you like to go out for dinner?" I said. I must have turned beet-red because I was feeling flushed from the spark of attraction and anxiety I felt.

Virginia put her hand on my arm and I relaxed. "Sure," she said. "I'm free tomorrow night." Soon, we were chatting like old friends and making plans for our first date.

Less than a year later, on July 1, 1972, a sunny Saturday afternoon in Scranton, Pennsylvania, Virginia and I said, "I do." We honeymooned at Ocean City, a beach on the New Jersey coast.

In our life together, Virginia and I have had adventures that took us from New Jersey to Gainesville to Atlanta to Hollywood and back to Jacksonville. While I've drummed in bands, worked in property management, and a little bit of everything, I'm in awe of this woman who made a career as a healthcare professional, and still had enough energy left to offer me her blistering spirit of love.

The death of another special girl, Lillian, no longer haunts my dreams. Vietnam took me to a dark place, but Virginia's brilliant sunshine lit my way back. I'll always be grateful for the K-9 heroes, for all MWD's, too. If it weren't for them (4000), so many of us would not have made it home.

I am blessed. Thank you, Creator!

Epilogue

Vincent DeGuilio lives in Napa, California and is a recently retired Sergeant with the Napa Police Dept. His two grown sons both wear badges, too.

Joseph Barbarise died of Agent Orange exposure from his time in Vietnam. I won't say much about this honorable man here because I'll do it later in the book I promised to write for Joe. He made me take pictures and write his story: the story of Agent Orange and our government's intentional poisoning of the troops and Vietnamese people.

Jimmie Moore with his dog Junior in Phu Cat, Vietnam, 1969.

About the Author

For me, the lives led by the canine handlers—that is, the U. S. Air Force Sentry Dog Handlers—and their noble canines, continued for twelve months (365 nights). Every day I counted the nights until my very own Freedom Bird arrived to take me home.

My dog Junior would be dead by December 1970—used up and discarded like all broken weapons. Four thousand noble canine gave all; they all gave all. This will never happen again; it must not happen again. It was so wasteful and disrespectful of all animal-kind. German shepherds were work-horses in Vietnam, and many of them were simply worked to death. Never again!

Jimmie Moore and his wife Virginia live in Florida.

To learn more about the author and noble canines,
visit www.mooreK9.com

The author with a new friend, 2016.

Made in United States
North Haven, CT
20 October 2026

20740138R00141